SAVING FARMLAND

SAVING FARMLAND
THE FIGHT FOR REAL FOOD

Nathalie Chambers

with Robin Alys Roberts and Sophie Wooding

Foreword by Nancy J. Turner

RMB

Rocky Mountain Books
www.rmbooks.com

Library and Archives Canada Cataloguing in Publication
Chambers, Nathalie, author
    Saving farmland : the fight for real food / Nathalie Chambers, Robin Alys Roberts, Sophie Wooding.

Includes bibliographical references and index.
Issued in print and electronic formats.

ISBN 978-1-77160-073-6 (pbk.).—ISBN 978-1-77160-074-3 (epub).—ISBN 978-1-77160-075-0 (pdf)

    1. Sustainable agriculture. 2. Sustainable living. 3. Farms, Small. 4. Local foods. I. Roberts, Robin Alys, author II. Wooding, Sophie, author III. Title.

S494.5.S86C52 2015   333.76'16   C2015-901004-7   C2015-901005-5

Cover Design: Chyla Cardinal
Interior Design: Frances Hunter
Front cover photo: Rural Scene © Karen Massier

Printed in Canada

Rocky Mountain Books acknowledges the financial support for its publishing program from the Government of Canada through the Canada Book Fund (CBF) and the Canada Council for the Arts, and from the province of British Columbia through the British Columbia Arts Council and the Book Publishing Tax Credit.

Canadian Patrimoine
Heritage canadien

Canada Council Conseil des Arts
for the Arts du Canada

BRITISH COLUMBIA
ARTS COUNCIL
Supported by the Province of British Columbia

This book was produced using FSC®-certified, acid-free paper, processed chlorine free and printed with vegetable-based inks.

CONTENTS

Foreword 7
Nancy J. Turner

Preface 11
Nathalie Chambers

Introduction 15

PART ONE Understand the Issues
ONE   Hear the Stories 21
TWO   Welcome the Bees 59
THREE   Taste the Connections 81
FOUR   Embrace Ecosystems 97
FIVE   Encourage Biodiversity 115
SIX   Transcend the Paradox 139

PART TWO Take Action
SEVEN   Overcome the Obstacles 147
EIGHT   Choose a Model 175
NINE   Identify Vital Farmland 201
TEN   Build Community 209
ELEVEN   Fundraise, Fundraise, Fundraise 225

PART THREE Keep It Up
TWELVE   Trust the Commons 251
THIRTEEN   Share Mutual Success 265
FOURTEEN   Keep Surging Forward 275

Acknowledgements 297
Select Bibliography 303

Nancy J. Turner

This is a hopeful book about resilient people who truly care about the well-being of the earth and of one another. It is at once informative and inspiring. It is also timely and important because our world is in peril. There are many signs that both cultural diversity and biological diversity are declining, as our human populations rise and as we co-opt more and more of the world's resources, taking over the habitats of other species and cutting them off from their food sources. Fossil-fuel production and consumption have increased exponentially, and this has led to undeniable changes in global climate: overall warmer regimes, with greater and more frequent extreme weather events. Many of us here in Canada feel that our food systems are secure and that we don't have to worry about getting enough to eat as long as the grocery stores are full of produce, no matter how it is obtained or where it comes from in the world. In fact, on Vancouver Island, we have only enough food collectively for about three days, should it stop being delivered from other places, and even now, many people are not getting enough to eat. There is a crisis looming, and it is, in fact, already upon us as we continually appropriate the best farmland for development and erode and damage already restricted food-production areas.

Saving Farmland is a testimony to the fact that we can turn this situation around, in a region like Vancouver Island and elsewhere in the world. For many people who read this book, especially those whose age is measured in multiple decades, its words and stories will resonate. For me, the descriptions of how farmland and natural spaces have been eroded through urban sprawl and other forms of development in southwestern British Columbia are all too clear.

When my family first moved to Victoria in 1953, and I was 5 years old, we bought an old brick house in Saanich, a block off Douglas Street. Across the street was a huge market garden, with rows and rows of rhubarb and vegetables, farmed by a Chinese-Canadian family. There was an open creek and wetland just a block away, and plenty of trees and shrubs all around. The B & K feed store was nearby, and behind the market garden was a large, south-facing garden owned by other neighbours, the Achesons. Every week I would walk up the block and around the corner to buy a dozen eggs from Mrs. Acheson, and in the summertime, when the pole beans and bush beans were ready in their garden, our whole family would go over to pick beans. Then my mother would get out her huge pressure cooker and our kitchen was turned into a bean-processing factory, with jars and jars of plump green beans being stowed away on shelves in a cool upstairs room, to be fetched and used in our winter meals.

Our own yard was quite large, and we had wild cherry trees along the road, with several apple and pear trees, cherries, prune plums, green gages, blackberries, strawberries, raspberries and

8

rows of cucumbers, tomatoes and potatoes, interspersed with my mother's favourite flowers: Canterbury bells, giant orange poppies, peonies, zinnias and rambling roses, among others. My parents kept honeybees in the back and the yard was always abuzz with bees. I remember there were always dozens of butterflies, too: coppers, blues, fritillaries, swallowtails, mourning cloaks and Lorquin's admirals. Bumblebees burrowed in the ground in some places, and mud-dauber wasps plastered their nests under the roof of our shed. My main entertainment around our yard was sitting in the backyard watching the birds – wrens, swallows, chipping sparrows, robins, towhees, juncos and finches, yellow warblers and goldfinches – build their nests and feed their young. Sometimes I would ride my bicycle farther afield to Rogers Farm, Swan Lake and Blenkinsop Lake, even where Madrona Farm is today. Always, for me, life in Saanich was full of interest and fascination. I was a part of the web of life, and, even as a child, I knew it.

My husband, Bob, had a similar experience where he grew up near the Tillicum-Burnside intersection. His parents and his grandmother, who lived across the back fence, had a huge garden, with corn, beans, tomatoes, onions and all kinds of other vegetables, king and Gravenstein apples, plums, cherries, blueberries, loganberries, raspberries and thornless blackberries. All of the produce was carefully harvested, and much of it was prepared for winter storage in the form of jams and jellies, preserves, pickles and applesauce. His grandparents grew loganberries in their backyard and sold them to the winery near Swan Lake. His family owned a grocery store and sold some of their home-grown produce to make ends meet. All over Victoria – and all over Vancouver Island and elsewhere – people were growing their own food, and valuing and nurturing the soil with compost and manure. Yards served as habitat

9

for many different creatures, and there was enough open space around that it was common to see garter snakes, lizards, bats, birds and diverse insects like golden-eyed lacewings, robber flies and June beetles.

The stories of what happened to all this diversity and to the opportunities for even city children to spend time in nature without going beyond their own backyards are well-known. "Development," "growth," "progress," "improvement": no matter how it is framed, it resulted in irreparable losses to biodiversity. It initiated the erosion of food production and severed our connections to the land and to the other species, even those on which we depend.

This is the situation that set the stage for the Madrona story, and for all the work that Nathalie and David, and so many others, have been carrying out. This is a joyful and inspiring story, a love story, in fact. We learn about people who form bonds with each other over common goals, about the deep and enduring connections to place, about selflessness and fortitude and positive energy. We see how some people can inspire others to follow their dreams, and how personal profit is not always the route to happiness. We see that even the seemingly *im*possible *can* be possible if people work in concert, if imaginations can be sparked, if we take on responsibility for caring for and protecting those things that *really* sustain us in this world. This book is uplifting, inspiring and hopeful. It combines stories about real heroes with profound lessons about energy, nutrient and water cycling, food webs, pollination and other natural processes. It leads by example, and it provides a template for preserving and enhancing farmland anywhere and everywhere.

From Global Grief to Action

Nathalie Chambers

The dream of saving Madrona Farm and the actions I took to realize that dream taught me important lessons that have changed my life and my position relative to the world around me.

Many of us feel as though we are drifting downstream toward a future where individuals lack effective power. It feels like the coordinates are set, that we can do nothing. The magnitude of this looming apparition affects each of us differently – some are more sensitive to the dread of this fate than others. However, we all feel it on a certain level: the thought of a powerless role in a future faced with environmental devastation, and all of the ills that come from it, hurts us physically, emotionally and spiritually. I call the symptoms of our collective despair global grief. Human health hangs in peril, as the building in which we are all tenants falls apart. However, as John F. Kennedy said, "Our problems are man-made, therefore they may be solved by man. No problem of human destiny is

11

beyond human beings."[1] It is a great paradox. We humans are the problem, but we can also be the solution. As individuals, we can work together for the common good; or we can be self-serving and destructive, living in fear.

We travel through stages of global grief: blame is a step on the way to acceptance, action and healing. We'd like to blame and issue eviction notices to those who we feel are speeding up our decline, for example, certain corporations, governments, people who drive cars, loggers, and exploiters of the tar sands. However, blame distances us from taking responsibility. With self-acknowledgement, honesty and humility, we can move on to acceptance; we are all to "blame." In the end, none of us is perfect, but I believe we all have a desire to be the best we can be and do something great for the world.

I remind myself constantly, "Every day you have a choice. You can choose to continue drifting along toward nowhere, or you can stand up and take action. You can choose not to live as a tenant in that crumbling building but instead become a manager, a steward. Yes, you can."

When I decided to become a steward, my first lesson consisted of three words: action dispels despair. Action soothed my suffering heart. I believe we are all hardwired for greatness; humans have a huge capacity for love, conservation, stewardship and peaceful coexistence. I would say to anyone in despair, "To spur yourself on to action, try to remember the dream to change the world that you had when you were a child. Let your loss of nature, those places you used to go to when you were a kid, the places that were essential to you becoming who you are – let those losses move you to act."

When I first walked on the land that is Madrona Farm on Vancouver Island, I made an unconscious connection. It felt as though the ground began to rule me. My direction in life changed, and some of my roots shot down deep into the soil. When Madrona Farm was threatened with redevelopment, I was moved to act. "Enough is enough!" I cried. "Stop the destruction!"

The Protect Madrona Farm campaign made me believe that change is possible. Any student of conservation biology will say that our problems are colossal, even insurmountable. I am an ordinary person who witnessed something extraordinary when 3,500 people in my community (and beyond) stood up to protect Madrona Farm in perpetuity during a grave economic recession in Canada, resisting wave after wave of seemingly impossible obstacles.

If we all work together, we can move mountains. When we as communities take a stand in our own unique places, we become a powerful force. When we stand together with our families and communities to protect nature, we are unstoppable.

We have everything we need in our communities right now to reclaim nature. As Margaret Mead so famously said, "Never doubt that a small group of thoughtful, committed citizens can change the world. Indeed, it's the only thing that ever has." During the three years of fundraising, sacrifice and strife that accompanied the Madrona Farm campaign, I couldn't wait until the issue was resolved. However, when one of our supporters, John Shields, senior program manager at The Q/The Zone radio station, asked me, "So, Nat, what are you going to do once this campaign is over?" and I said I would get back to farming, he answered, "Yeah, right. I know people like you. They just never stop."

He was right. A thought occurred to me a week or so after the campaign was over. "Okay, Madrona is protected – but what about all the other farms?"

In 2013 I went to the International Conference of National Trusts in Uganda, where I presented the Madrona Farm conservation model. I came home from the conference in October that year to discover two things: The Land Conservancy I had just been touting as "the mechanism to protect farmland forever" had entered into creditor protection; and the British Columbia government was threatening to dismantle the province's Agricultural Land Reserve (ALR), a system that, up until the end of the BC Legislature's spring session, had reserved about 4.7 million hectares (116,139,529 ac) in the province for agriculture. The ALR, of course, was an obstacle to fracking and oil and gas development. (As of this writing, 90 per cent of the ALR is threatened with redevelopment.)

Overwhelmed by the news, I turned to a community group, the Social Environmental Alliance. Together, we formed the Farmland Protection Coalition to address the province's move to overhaul the ALR. Our group held a Family Day Rally at the Legislature in Victoria that was attended by thousands. We helped to facilitate six town-hall meetings in the province. We built a contact list of over a thousand, and more than 65 organizations endorsed us. Instead of relying on The Land Conservancy alone, we realized we had to change our strategy for saving future farmland by creating more all-encompassing protections through a variety of local, municipal and regional land trusts that work together in partnership. We are now working on creating local, municipal and regional land trusts working in partnership with other land trusts.

I now understand that nothing is "forever." There will always be obstacles to conservation, so a conservationist's work is never complete. This book is about a part of the greater conservation picture: the process of action required – the quest – to protect farmland for future generations.

INTRODUCTION

Too often, we feel powerless to change the powers that be. To step outside our own marginalization, we need to look at the world through a different lens. If we think of American Farmland Trust's "Earth as four slices of an apple," how thoughtfully can we share it?[2]

Three slices of the Earth apple represent oceans, lakes and rivers. The fourth slice represents land. Half that land – 1/8 – is too hostile to support us: too hot, too dry, too cold or too rocky. That leaves the other 1/8 for us to live on. Seventy-five per cent of that dwindling slice – 3/32 – is either covered with soil too poor for agriculture or smothered by cities, towns, highways and byways. Some of the other 1/32 might be employed to grow food. However, to find the real dirt on our Earth apple, we must pare that last slice down to its peel, where we find the Class 1 and 2 topsoils optimal for growing food.

In the last ten years, investors have bought up enough farmland internationally to cover the UK eight times over – either to speculate on real estate or to grow biofuels.[3] Every year for the past 25 years, the US has lost almost 404,686 ha (1 million ac) of farmland.[4] In 2012 the UN reported that food insecurity is on the rise in Canada; the number of food insecure people in Canada rose from 1.92 million in 2007 to 4.3 million in 2011.[5] Earth has to feed 7.1 billion people, yet one out of every seven goes hungry.

In 2011 $868 would feed the average British Columbia family of four for one month – although after paying rent, those families relying on income assistance were left with less than half the amount needed for this average food budget.[6] Studies have shown that residents in Vancouver homeless shelters who are fed three full meals per day have a much lower tendency to be violent and an increased tendency to resolve their housing problems than those who have less to eat.[7] Because the potential for violence exists at these shelters, women avoid going to them, even though meals are more dependably available there and consist of fresher foods than those available at food banks.[8] Yet, single mothers and their children are eight times more likely to suffer from hunger than any other family type in Canada.[9]

It's clear we need more food, but over the last 20 years, the number of Canadian farmers under the age of 35 has plummeted from 77,000 to barely over 24,000. Farmers are advising their children to avoid the stress and debt they have experienced in their vocation.[10] However, the single biggest factor blocking food security is the price of farmland. The price of agricultural properties located close to growing cities has the potential to equal that of residential real estate, if its zoning can be changed.[11] Preserving arable land close to city populations has benefits for both farmers and city residents. Understandably (but sadly) finances often propel income-stretched, tired and older farmers to wade through extensive rezoning campaigns and paperwork, resulting in the fragmentation of more good food land. Opportunities for young farmers continue to decrease as the amount of arable land shrinks and the amount of permanent cropland (think fruit and nut trees, as well as grapevines) dwindles. Available arable land in Canada dropped from 5.0 per cent in 2001 to 4.73 per cent in 2011. Permanent cropland in Canada tumbled from 0.7 per cent to 0.5 per cent over the same period.[12]

Despite the promise of the questionably named Green Revolution[13] and its heirs – companies like Monsanto, with its GE technology and GM seeds – industrial agriculture is failing to feed the world. Why? We've known for over a decade that the costs of producing food and the land on which to produce it are increasingly at odds with the tighter and tighter concentration and distribution of economic power.[14] As we face the removal and loss of biodiversity, pollinator decline, salinization, desertification, acidification and soil degradation, traditional sustainable agriculture practices are on the brink of collapse.

As Nathalie Chambers mentions in her preface to this book, we each have the power to move mountains. If the collective impact of Madrona Farm's community can succeed during one of Canada's worst recessions, others around the world can take it upon themselves to save even more farmland. Save it while we still have the capacity to seed perhaps a bit more than 1/32nd of that metaphorical apple peel; save it to represent the kind of landscape we want for future generations. Nathalie wants to clearly illustrate what this future could look like so that people can climb out of bad-news paralysis and make a difference. She feels she is not special in this regard; she knows that everybody can make a difference, especially when we stand together.

After Madrona's successful campaign, Nathalie felt as though she was hanging in a pink cloud. She felt as though no problem was too big for the community to take on. She began scouring the infinite possibilities for projects, knowing that we collectively had the solutions. Refusing to simply plop back down to earth without sharing her lessons, she asked Robin and Sophie to help her write this book. Throughout her life as a farmer, Nathalie has learned that farming and agriculture encompass everything: culture, heritage, food, people, trees, flowers, frogs and bees. As she studied agroecology, she understood that it all grows out of one pot. Agriculture conservation became her pot. Thousands of food-loving people have already jumped into it. Together, we wrote this book to share both Madrona's model and the model of international conservation that is now blossoming in its purpose to put food first.

Welcome to *Saving Farmland*.

PART ONE
Understand the Issues

> "In the long view, no nation is healthier
> than its children, nor more prosperous
> than its farmers." —HARRY S. TRUMAN[15]

ONE : Hear the Stories

The Stephens: Paradise Gained and Lost

Paradise

In the cold dawn of January 6, 1944, Gwen Stephens heaved
sacks of potatoes from the back of her family's old Ford truck.
Before noon, the dark green pickup was bouncing her ten
miles (16 km) north of their Mountain Valley farm to King's
Daughters' Hospital in Duncan, BC, where she gave birth to her
third son, Arran. A Scottish isle inspired his name.

Arran says of his dad, Rupert, "While not formally religious,
he taught me the sanctity of the forest, that it was our church.
And I did find my sermons in stones, and books in rivulets."[16]
As signs of their presence, deer left pointy hoof prints, nibbled-
off leaves, shorn grass, rounded scat, scuffed tree trunks and
torn branches, even the occasional old antler along the weaving
paths of the forest hugging their homestead.[17] Trout leapt in the
streams, babbling their tales to a boy who cared to listen.

"Mum loved the feel of the earth," he said, "the subtle patterns in a stone, the colour of a leaf, a tiny wildflower springing from emerald moss, water dancing in sunlight, maidenhair ferns waving on a wet canyon wall." Whether she was filling his head with the sounds of the common and Latin names of local wildflowers or tantalizing his senses with the smell of freshly baked cookies, which they took to the abandoned elderly, Arran grew up steeped in the values of connection and compassion.

Rupert built two large rhubarb houses that he covered with earth so they could be the first to offer rhubarb early in the spring. He developed rare strains of berries, and his mulch kept their plants' toes warm enough that the family could sell everbearing strawberries as far into the autumn as November. He sold nursery stock and generously shared his inventive farming methods. "Great teachers, they were," says Arran of his parents, "full of life and wonder."

When Arran was 6, his parents decided to sell the family farm, even though his granddad's tombstone lay nearby. Queen Victoria had awarded Dr. Harold Stephens this wilderness acreage for his exceptional service as a naval fleet commander and surgeon in India. But better land beckoned 17 miles (27 km) south in the Goldstream Hills near Victoria. Toppling second-growth forest, levelling two large terraces to create back fields and building a log house with a view over the valley, they created their dream: Goldstream Berry Paradise. Swaths of yellow forsythia and Scotch broom lit up the steep slopes. With storage sheds and an enticing roadside stand by the highway, they sold their fresh produce alongside Babe's Island Honey.

With his dad, Arran planted corn by hand. In the fall when seaweed washed up on the beaches, the family drove south to Sooke to load the back of their pickup. After dragging long tubes of kelp between the rows of strawberries, they'd swing their machete to chop it all up, to help it rot faster over winter. As the worms processed the mulch and enriched the soil, Rupert told his son, "I call him not a man who knowingly steps on a worm."

Contradicting conventional wisdom, Rupert decided to cover their fields with chicken manure topped by a four- to six-inch (10- to 15-cm) layer of sawdust. Squeezing out weeds and keeping in moisture, his experiments saved both work and water. Neighbouring farmers were so surprised by the Stephens's impressive yields – more than ten tons of berries per acre (22.4 tonnes per hectare) – that Rupert felt encouraged to write about his work. His treatise, *Sawdust Is My Slave*, became a popular guide for using mulch to create a paradise for earthworms.[18] One of Rupert's maxims remains a family mantra to this day: "Always leave the soil better than you found it."

Wherever Arran hiked on the farm, his cat, Dizzy, skipped after him. "From the time he was a fuzzy little kitten," Arran said, "I had trained him to turn like a whirling dervish whenever I circled my finger in front of his face."

Paradise Lost

When Arran was 13, Rupert decided to sell the farm and pursue his dreams as a songwriter. As Arran gazed for the last time at their log house and farm nestled in the blue-green mountains, he said, "I wept uncontrollably while Dizzy yowled in the field, vainly trying to follow our car as it rolled down the tree-lined Humpback Road for the last time. Beloved Dizzy was left to fend alone on the abandoned farm. Soon, I too would have to learn to survive, far away in the concrete jungles of L.A., San Francisco and New York City. "

Leaving the fresh country air, they crossed the Straits of

Juan de Fuca by ferry, climbed aboard a train in Port Angeles and headed south through Washington, Oregon and most of California. In smoggy Hollywood Hills, the family moved into a rented hacienda shaded by palms. Flung from the freedom and wilderness of southern Vancouver Island, Arran said, "I learned to my dismay that violent gangs controlled the schools and streets, where the wary and the strong survived. Not particularly large or strong, I learned a little Okinawan jiu-jitsu for self-defence, which taught me that even someone small or weak could use the momentum of someone much larger and stronger against them. In Grade 8, I had only taken a few lessons before becoming the unprovoked object of attack by the toughest gang leader in Le Conte Jr. High, a very tough school on the edge of East L.A. When the bully ended up unconscious on the ground without my having to strike a blow, I sensed a hidden life force, called chi'i, or prana. After this unusual experience, the superstitious gang left me alone, to my great relief."[19]

But "one hot afternoon as I was watering the plants," Arran adds, "a lean, orange-white tomcat walked up the front steps toward me. Beneath the grime was a remarkable resemblance to my long-lost friend. 'Dizzy, is that you?' I whispered, circling my finger in front of his face. To my utter amazement, that scraggly cat started to turn around in circles. Next, he was purring and blissfully rubbing my leg. Sweeping him into my arms, I ran to the house, yelling, 'Look, look, it's Dizzy!' Somehow, he had crossed the Juan de Fuca Strait, and travelled 1,500 miles overland to find his boy in the middle of a city of millions. What a story Dizzy could have told! Who says animals don't have souls?"[20]

During the next ten years, Arran crossed continents both internally (writing poetry, painting, practising monastic meditation) and externally (from California to New York to

India). Like his grandfather, Arran married a remarkable Indian woman. He and Ratana met in an ashram and within weeks, decided – with the blessing of their meditation teacher – to be married. Returning to western Canada months later, Arran was eager to show his cherished Goldstream Berry Paradise to Ratana.

Near the location of the old farmstands, a shabby trailer park greeted them. As Arran said, "the forgotten farm fields and the big hill where our big log house commanded the Goldstream-Humpback valley – gone! The entire hill was removed for gravel extraction. The log house vanished, all the fertile fields covered with houses, chemically fertilized lawns, asphalt roads and driveways."

They drove up to his grandfather's old Mountain Valley Farm "only to find the surrounding mountains all logged, and the clear trout stream choked with stumps, rubbish, brush and silt," said Arran.

Staring at the lopsided, crew-cut forests, the filthy watershed and the grungy mobile homes where once he'd danced on a farmstead surrounded by nourishing forests, the home buried deep in Arran's heart pounded with his dad's words. *Always leave the soil better than you found it.*

"Aagggh!" he said. "I never wanted to see it again."

Harold and Kathy Steves: Tenacity Surrounded by Loss

Choosing the Wetlands

In the fall of 1877, Manoah Steves stared silently at southern BC's marshy Fraser River Delta, contemplating his past farms and how they related to his future life. He, his wife, Martha, and their six children – ranging in age from 2 to 20 – were all born in the wetlands of Moncton, New Brunswick. Over the previous decade, he had moved his family to Chatham, Ontario, where they farmed near the shores of Lake Erie, and then to a peach farm in Maryland. After a malaria attack on the humid shores of Chesapeake Bay, Manoah set off in search of a healthier climate, promising to send for his family as soon as possible.

Vancouver's Coast Mountains stood tall at his back. Washington State's Mount Baker rose in white elegance to the south. Moncton lay 4,040 miles (6,500 km) away, nestled on an elbow of the Petitcodiac River bounded by Lutes Mountain to the north and the jagged Caledonia Highlands to the south. Manoah finally felt at home.

Even though part of his new farmland on Lulu Island would disappear under water at high tide, he promptly bought four hundred acres (162 ha) at 75 cents an acre. He understood the marshland: how to dyke it, drain it and farm it. As he squeezed the alluvial soil between his fingers, he felt its dark, fine-grained, fertile richness and gave it his highest rating. The Steveses would become the area's first white family of settlers.

Family Cheer

On the morning of May 24, 1878, the sidewheeler *Enterprise* steamed over from Victoria, carrying Martha and the children. After steering inland from the Salish Sea around Garry Point, the boat began heading up the Fraser River, eight miles (14 km) to Ladner's Landing and another 15 miles (25 km) to New Westminster. Suddenly, the family noticed their father waving from a huge log at the river's entrance. As his three older children (Herbert, Josephine and Joseph) cheered and waved, the three youngsters (Mary Alice, Ida and Walter) jumped up and down on the deck. After earnest discussions, Martha managed to convince the captain to stop midstream. Lowering a rowboat and carrying the younger children down on their backs, the crew ferried the entire family and their possessions to shore.

Within a century, Garry Point would look quite barren, but during the Steveses' boisterous reunion in 1878, the sweet perfume of wild rose bushes wafted around them. Beyond the solitary beach log and the thick roses, tall grass filled the few spaces between spruce, cedar and a variety of deciduous trees. The top of a man's head suddenly appeared above all the foliage. Generosity, in the shape of Garry Point's only other resident, six-foot-four (1.9-m), elderly bachelor Edward Sharpe had arrived.

Edward helped the family haul their luggage three-quarters of a mile (1.2 km) inland along wild cattle trails to his 12 by 12-foot (3.7 × 3.7-m) shack, which was built above the marsh on driftwood-board stilts. Inviting the family up the curved, rickety stairs to his bedroom-sized home, Edward pulled the latch string on his door and welcomed the family inside. Until they could build their new home, Martha, the three daughters and young Walter lived in Edward's shack while Manoah and the older boys slept in the barn with Edward.

The Steveses designed their house to be almost twice as large as Edward's shack. With a barn-style door opening into the downstairs kitchen and living room, it had two bedrooms upstairs. First, they pounded six-foot (1.8-m) pipes into the ground to position the dwelling above May and June's flooding high tides. After planking its walls, the Steveses imported real glass window panes by steamboat so they could look over the earthen dike that they built to safeguard their backyard.

The family could use the abundant river water for washing dishes, clothes and floors, but it was delta water, too salty to be potable. The thick clay soil offered no hope of well water, either. Even almost 20 years later, when borers put down a pipe first to a depth of seven hundred feet (213 m) for a dollar a foot ($1/0.3 m), then at a further rate of two dollars a foot ($2/0.3 m) down to a total of 1,008 feet (307 m), they found no fresh water.[21] Instead, the Steveses built barrels to collect rainwater. However, before drinking it, they had to filter out hordes of wiggling insects. During the dry season, they rowed out and scooped up the relatively untainted water from nearby mountain streams that flowed through the river's centre.

Manoah fashioned his own fishing net by tying a series of string knots. He caught plenty of salmon, which the family ate fresh and also salted for the winter. With his eight-gauge shotgun, Manoah also killed up to 150 birds at a time, which the girls beheaded, de-winged, skinned and salted. The boys dug clams and cockles at low tide and dip-netted for eulachons, while the girls gathered crab apples, salmonberries and gooseberries for jam and preserves. Native people sold the family candlefish and sturgeon, one of which could weigh as much as 1,200 pounds (544 kg). (Manoah told his family that big sturgeons frequently ruined nets. When some fishermen finally untangled one sturgeon from a net, they heaved it into a three-ton milk truck with its tail tied to the roof. After chauffeuring it to Vancouver, they exported part of it to New York.[22])

Growing the Steveston Farm

Although Manoah, Herbert and Joseph used a horse and plow to try to clear a small patch of land, the ground was too boggy. Borrowing Edward Sharpe's team of oxen, they managed to plant some hardy vegetables and a late batch of oats, which had to be harvested green from the salty soil for winter fodder. Because the ground was too wet for the usual root cellar, they eventually stockpiled potatoes, turnips and cabbages in a mud enclosure above ground. That first batch of green oats became fodder for a cow rented from a neighbour. Soon after that cow calved, the Steveses purchased a second calf from another settler. One day, they walked into their garden to be greeted by a third calf, which had wandered through their fence. In 1886 Manoah imported BC's first purebred Holsteins from Ontario and New York. The herd gradually grew to two hundred head. Manoah also grafted ten acres (4 ha) of wild crab-apple trees with stock from standard apples, and he imported several varieties of pears. His first fruit cheque, which came at the end of the 19th century, amounted to 50 cents.

During their first decade in the Richmond area, all of the area's new settlers laid out walking trails on top of the boggy terrain, but they quickly grew overwhelmed trying to slash through the delta's thick underbrush. Rowboats, steam tugs and ferries provided their main modes of travel for shopping, attending church, visiting friends or going to meetings. If they did manage to walk on Lulu Island, they always wore gumboots and carried slippers wherever they went. Off-islanders soon dubbed the new islanders "mud flatters."

The nearest store was in New Westminster, 17 miles (27 km) upriver, so Herbert traded the new double-barrelled shotgun he'd brought from the east for a rowboat, and the Steveses rowed the 34-mile (55-km) round trip for staples – flour, salt, rice, coal oil, candles and soap. A typical visit to their nearest neighbours or the Ladner post office meant a five- or six-mile (8- or 9.5-km) trip. For the first mile, they'd row east against a headwind for an hour. Then they'd wait at a fish camp until the fishing steamboat arrived and could tow them the rest of the distance.

Within a couple of years of the family's arrival, salting gave way to canning fish, and Herbert bought his own land on Lulu Island, thus giving birth to the town site of Steves, later called Steveston. In late 1885, the Canadian Pacific Railway finally reached Vancouver, which made it easy for both Herbert and Manoah's second-oldest son, Joseph, to establish western Canada's first mail-order seed business in 1888. By 1901 Steveston had succeeded so well with its 15 waterfront canneries supporting the fisheries and boatbuilding industries that the community sent a record 16 million pounds (7,257,478 kg) of salmon in every direction, including overseas. Locals soon nicknamed it "Salmonopolis."[23]

Joseph Steves inherited his father's passion for breeding dairy cows and expanding and improving milk production. By the early 1900s, when the family had built up the dairy to the point that they were shipping ten-gallon cans to Vancouver, they arranged for the empty cans to be filled with drinking water on the return, which provided a much tastier alternative to rain-barrel water. (By this point, barrels graced the outside walls of every house on the island.[24]) Eventually, Joseph's dairy provided stock for the entire province of BC.

Meanwhile, Manoah spent increasing amounts of time sitting on the council of the newly formed Municipality of Richmond. In his 1927 *History of Lulu Island*, Thomas Kidd describes Manoah at age 52, "still a vigorous man with hair and beard showing grey, a man somewhat reserved in conversation, but whose rather wide experience made his advice valuable in municipal matters, and whose hopefulness of Richmond was unbounded."[25]

At the time, little velvet-leaf blueberries, wild cranberries and huckleberries contributed to everyone's diet. Following the practice of the ancient Native peoples, the hard-working Chinese, who had originally immigrated to work on the railway, lit fires on sections of the bog. This quickly rid the land of dead canes, leaves and stems that could cause disease in new growth and gave room for the berry bushes to sprout up in vibrant health in the rich peat.

In 1899 Joseph's wife gave birth to Harold L. Steves (Sr.). As a young fellow, he remembers walking "on 8-foot board sidewalks over the ditches in front of the stores." Because the bog was so wet, parents advised their children to stay off it. Although at first glance it did not appear dangerous, in some places, a person could fall as far as 16 feet (4.9 m) straight down.[26] After losing some of their horses to the bog, farmers fitted the front feet of their horses with specially designed "tule shoes," at first made of eight by ten-inch (20 × 25-cm) wooden planks, and later fashioned of 12-inch (30-cm) iron rings attached outside the metal horseshoes. The animals had to learn to swing their shoes outward as they walked to avoid tripping.[27] When curious visitors arrived from Vancouver, locals considered it a "Sunday sport" to heave their cars out of the mud.[28] Because of the treachery of the bogland, Harold explains, "The road was planked, too. After fishing season closed, if you wanted to go down on a Saturday night, you'd have to walk down the road as there were too many people on the sidewalk."[29]

Harold remembers blinking as the first electric lights turned on, and soon after – in 1906 – he experienced the arrival of the first electric trams. When electric power arrived for homeowners in Steveston in 1920, they had to install their own pole and lines. Joseph bought one of the first electric motors in the delta area for the family's barn.

Harold Sr. began school in Steveston and despite a bout with typhoid fever, graduated from Richmond's first high school. Transportation to and from the island continued to be challenging in those years. In 1917 he lived on campus for his first year at the University of British Columbia, which (until 1925) was located in the Fairview area of Vancouver. After a two-year break, he spent the subsequent years until his Bachelor of Science in Agriculture degree in 1924 commuting 16.9 miles (27.2 km) on the Canadian Pacific Railroad's electric train. Salmonopolis locals called this twice-daily commuter the *Sockeye Limited*, as it catered to Steveston's busy fish canneries as well as the Lulu Island farms. Passengers said the automatic reflex of their noses to the strong smell of fish signalled their approach to Steveston.[30]

Until about 1920, when Ford vehicles sporadically rumbled along the gravel or wooden roadways, most island farmers used horses and mules for transportation. They brought their roadway gravel from ships that dumped their ballast nearby while scows supplied crushed rock for further cover. On top of the dyke, the island community also built a raised pedestrian walkway made of three-foot (1-m) -wide planks. This boardwalk allowed people to walk all the way around the island.

In the delta's early farming days, a natural barrier of centuries' worth of logs and rushes kept the water out. One day, some pigs rooted a hole through this natural heap of debris, and water poured in. Farmers often faced the dangers of breaking dykes. After a full day's shift on the farm, they'd spend two to 11 hours each night patching up the holes caused by pounding logs during storms at high tides. The most serious run-offs occurred during sudden changes in temperature from cold to warm – one in 1894 before the dykes (when five feet [1.5 m] of water poured over their farms) and another through the dykes in 1948.

Early in the century, the Steveses had built up a herd of 70 or 80 Suffolk Punch workhorses. These horses pulled wagons and buggies, cultivated the gardens and tilled up the root crops, but they avoided most heavy work. The Steveses kept two big stallions for breeding; everyone needed a saddle horse, and many of the farmers loved horse racing. Originally from England, the family's workhorses won the 1909 World's Fair championships in Seattle. Sadly, when a contagious disease called glanders[31] swept the world nine years later, and the government tested all North American herds, the Steves had to accept the extermination of about a quarter of their herd.

While working in the delta soil, the Steveses continually brought up buried logs. One day in 1918, their plough became stuck trying to haul out an unusually big log, measuring ten or 12 feet (3 or 3.6 m) in diameter. They sawed the log in half but still couldn't budge it. Finally, they employed a powderman specialist miner. Bringing blasting powder from a Vancouver hardware store, he inserted eight sticks into the log and detonated a hole as big as a bedroom. The contagious workhorses – which had only just been shot – were buried in that hole.

The giant log's firewood provided steam for their threshing engines for the next three years. They mounted their first steam engine on wheels. Their four-horse team towed it into position. The steam engine turned a flywheel that moved a long belt to power their two separators. As the men pitchforked the sheaves onto one, the belt powered the removal of the straw. The other thresher separated the grain from the chaff. The family then baled their hay and bagged their grain.

By the turn of the 20th century, the Fraser River Delta had grown into a hub of farming, especially at the delta's ocean front near the Strait of Georgia, on Westham Island – which was intensively farmed. Steveston, on the much larger Lulu Island shore, became the delta's centre of activity. Ladner, on the southern shore of the Fraser River, housed a few fishermen. The Fraser River rarely froze, but in 1894 the weather was so cold that people were able to haul produce across the river's two iced-up arms. They even navigated the 18 miles (30 km) of river on ice from Steveston northeast all the way to New Westminster.

Fishermen helped strengthen the chicken farmers' eggs by selling hundred-pound (45-kg) weights of crushed clam and oyster shells, which they hauled by sailboat from Vancouver Island to Lulu Island. Limestone in the shells provides the calcium that all hens need to produce sturdy, thick-shelled eggs. Although free-range birds can ingest a measure of calcium and phosphorus from eating hard-shelled beetles, crushed oyster and clam shells provide a more concentrated source of the minerals.

As farming and fishing activities became more and more established, Lulu Island residents finally finished the road by building it up high enough over the bog and levelling it off. A few of the early airplanes began landing on the old racehorse track – even racing alongside the horses. Harold said one horse beat the plane, but the pilot had given the horse too much of a head start.

As improved transportation eased access on and off the island, the Steves family expanded their farming activities. While continuing to nourish their dairy of grass-fed Holstein cattle, which provided milk for many in Vancouver, Harold and Maude Steves also became western Canada's first heritage seed protectors and collectors. In 1937 Maude gave birth to Harold Steves Jr., who, like his father, graduated from the University of British Columbia with a degree in agriculture. He specialized in genetics because, as he said in the summer of 2014, "Our family established the first seed company in BC in 1888. We are still growing, saving, improving, and selling seed from over 50 vegetable varieties adapted to the BC climate."[32]

Sadly, Harold's graduating class at University of British Columbia turned out to be the last one at that institution to learn the skills associated with organic farming.

The New Barn Challenge

By the 1950s, the Steves family needed a new barn. Harold Sr. went to Richmond's city council to apply for the necessary building permit. Recalling the incident more than 50 years later, Harold Steves Jr. said, "It seems like yesterday. I'd just milked the cows and come in for breakfast. That's when Dad gave us the news."[33] The municipality had made a silent zoning change: the land had been rezoned for residential subdivisions. No new barns would be allowed. Harold felt the passion of three generations of family farming – of the agricultural history of the delta – rise above the boiling point as he tried to grasp the capricious ignorance of the city councillors flipping the lower mainland's central food source into housing.

At a big meeting in downtown Richmond, many protested, but the municipal government ignored them all. The zoning went through. One by one, the 12,000 acres (4856 ha) of farms on Lulu Island – some of the most prime farmland in all of British Columbia – faced extinction. What would father and son do next?

The Making of Madrona's Farmstand

Ancient History and Food Practices

While tending Madrona Farm from the turn of this century, Nathalie and David Chambers have grown with the lessons from its soil, from its neighbours and from its deep past. They understand the necessity to dig into its rich history. As the Chambers restore the natural habitat within and around the farm, they realize the possibilities for returning Earth to its beautiful, sustainable potential while producing healthy food.

Madrona Farm lies on the western slopes of PKols – or

"White Head" – as Mount Douglas was originally called in the Native language. Located on traditional lands that the Saanich people have stewarded since their ancestors gathered around the edges of melting glaciers, PKols held special significance as a meeting place. From its 213-metre (700-ft) head, different families and nations exchanged news, and viewed the weather, the surrounding waters and the people paddling home from nearby islands. They hunted and roamed freely. In those early days, Haro Strait on the eastern side of PKols connected with Blenkinsop Lake on its western foot.

Gradually the sea receded, revealing Blenkinsop Valley, graced with fertile soil. For thousands of years, Native peoples thrived in the area by listening to their Elders. They learned how to identify and treat each ecosystem with respect. They understood how to maximize the available diversity. They forecast the best window for harvesting and coordinated other seasonal activities. Although they manipulated plants and their environments to promote the growth of culturally preferred specimens, the people also worked with care to enhance specific plant communities. "Keeping it living" was their mantra.[34]

Knowing which bulbs were edible and which poisonous, the Native peoples on the island managed camas plots as sustainable farmers do.[35] For more than two thousand years, they tended berry plots. "Burning specialists" carefully timed burns to produce more and better blueberries or blackberries. Such burns ensured reproduction, thus increasing yields of edible roots in Native agricultural patches up and down the coast.[36] They cared for hundreds of species of plants, from red cedars on hillsides to tiny bog plants in river estuaries. They understood the immense mulching and restorative benefits of the salmon runs. They nourished the soil, using special digging sticks to aerate their productive farmlands.[37] They welcomed the wild rosehips after the first frost. They calculated when their favourite berries would be ready to pick and when it was time to greet different birds as they flew back from migration. They understood the animals skirting the bases of their favourite trees, and they knew how and why to preserve those same trees, as well as the ponds, lakes and rivers. The Native peoples traded information amongst their many cultures, so that over the centuries, their cedar baskets filled not only with food but also with current and ancient knowledge of how to support the habitat that so fully supported them.[38]

The Pacific Coast Native perspective on agricultural development recognized that the creation of resources was already complete. With nature already in abundance – providing constantly appreciated wealth – humans "merely tend, nurture, and respectfully, thankfully, take what is needed. Where people seek balance and harmony as a way of life, the earth and its resources become equal partners in an endless cycle of respectful life."[39] The Native model, which thrived for thousands of years prior to European Contact, is diametrically opposed to current culture, "which demands that humans dominate the environment and profit from it."[40]

The New Settlers' Impact

Until a little over a century and a half ago, Native cultures in the vicinity of modern-day Madrona Farm thrived, as their deep understanding of the lavish land wove through every part of their society. Then European and American people arrived to "settle" the land. In 1850 a series of treaties was negotiated by Vancouver Island's governor James Douglas, with local peoples and signed by leaders of the Songhees Nation, wherein Native leaders transferred land title over to the Hudson's Bay Company. Called the Douglas Treaties, they involved 14 land title purchases, with land exchanged for money, clothing and blankets.[41] They formed the only treaties made with Aboriginal peoples in British Columbia for nearly 150 years, until the Nisga'a Treaty in 2000. Perhaps the Songhees Nation signed the Douglas Treaties because it had been assured the people would be allowed to "roam freely" over the traditional lands. The Saanich Peninsula cost the company £109, seven shillings and sixpence for the approximate total of 386 wool blankets, which the various Native leaders preferred to receive instead of cash.[42] However, as University of British Columbia researcher Wilson Duff mentions:

> Why did Douglas not sign the treaties himself . . . ? As agent of the Company and the Crown, he was one of the parties to the transactions. He took pains to have the other party (the Indians) make their "signatures" and had his employees witness them, but nowhere did he affix his own signature. Why are all the Indian X marks so regular? It is not because the documents are copies, for it is clear from other indications that they are the originals. Could the answer lie in the manner in which the marks were made, perhaps with the Indian just placing his hand on the pen as Douglas made the mark?[43]

As the settlers took more and more ground, introduced alcohol and imposed a new religion, they ripped the foundations of Native cultures apart. Rather than listening with respect to honour lessons they could have learned about how best to sustain both themselves and the land, the new people chose only to force the Natives to abandon caring for the whole area and to confine themselves to selected "Indian Reserves."[44]

When the Hudson's Bay Company logged PKols to build Fort Victoria in 1843, it destroyed the majority of enormous native cedars that had prompted the colonials to bestow the little mountain with the name Cedar Hill. Douglas firs grew up to overshadow the remaining smaller cedars. However, in

38

1889, the white settlers made a pronouncement. Henceforth, this site, sacred to the local First Peoples, would be preserved as a park – but it would be renamed Mount Douglas after Queen Victoria's representative, Governor Sir James Douglas.

The newly christened Mount Douglas quickly lost its fertility. When its slopes eroded – first due to logging, then as a result of three major fires followed by rain – the Native peoples had to rebury the bones of their ancestors. White settlers – not communicating with Native peoples and not seeking to understand their surroundings – also destroyed a Native midden at the foot of the mountain, a place that had long been an important all-tribes meeting centre. Paddlers used to come to what is now called Mount Douglas Park beach from all over Saanich and the San Juan Islands for clam bakes and cultural exchanges. After such affronts to the mountain's significance to Aboriginal culture, recent efforts point toward reclaiming it. In the spring of 2013, Tsawout First Nation chief Erik Pelkey led a march up to the mountain's peak to share stories and rekindle awareness of its origin.[45]

Despite their ignorant approach to Aboriginal knowledge, sacred sites and culture, white settlers in the mid-1800s did work valiantly within the limitations of their European cultural understanding to create a productive food base on the land they had taken over. Farmers grow food on this southern tip of Vancouver Island 12 months a year because the area is in a perfect rain shadow, surrounded on three sides by Washington's Olympic and Cascade Mountains, Vancouver's Coast Mountains and the Vancouver Island Ranges.[46] Victoria is situated in a climate much like that of the Mediterranean, having only 92.7 centimetres (36.5 in) of annual rainfall compared to Vancouver's 145.5 (57.3) (or North Vancouver's 252.2 [99.3]).[47] The land around Mount Douglas also consists of nine different types of soil, including white beach sand, which provides a rich mineral base for top-quality produce. A biodiversity hot spot, the area around Mount Douglas is part of the Garry oak and associated ecosystems – hosting 137 different pollinators. Food lovers find all of this within a 12-minute drive from downtown Victoria.

How the Chamberses Found Their Dream Farm

The first farm on the hillside sprouted in 1853, when newly planted Douglas firs emerged to replace the cropped cedars. As more and more farms grew, they began feeding both locals and other islanders. A century later, Ruth and Lawrence Chambers bought Madrona Farm on the western slopes of PKols/Mount Douglas Park.

Ruth's father, Max Enke, had certain characteristics and skills – a will to survive, intuition, creativity and generosity – that he passed on to his daughter. An Englishman who immigrated to Galiano Island in the 1920s, where he bought 1,200 acres (486 ha) on its southern shores, Enke also held family property in Belgium. In the 1930s, he returned to Europe for a few years to tend to that property and business interests.[48] Early during the Second World War, however, the Nazis captured him, and he endured five years in a concentration camp. During three of the years he was incarcerated, he sent letters via Red Cross mail drops to start fulfilling his dream to reserve part of his Galiano property as a park. When the first plane arrived to take camp survivors home, he refused to go. As he watched others racing to climb aboard, Enke's intuition seemed to speak for him as he declared, "I've been here for five years. I'm in no hurry now." That plane crashed. Eventually, of course, Max did find his way home.

Three years after the end of the war, Max and Marion Enke completed his dream and donated over a quarter of their property – 317 acres (128 ha) – to Galiano Island. Called Bluffs Park, it includes virgin forests, high cliffs with arbutus trees gracing southward views overlooking Active Pass, and a long sandy beach.[49] Thanks to Max Enke's concern that nobody should own that property, he arranged that it be kept for public enjoyment, in perpetuity.

Ruth's husband, Lawrence, spent his early childhood in England and northern Alberta. In his young teens during the Depression, he went to Vancouver to find work. The hunger he experienced during that time inspired an insatiable desire to grow food. When he married Ruth and began raising their three sons in Ladysmith, she encouraged him to fulfill his dream. In 1952 they drove south 56 miles (90 km) to Victoria, searching for a farm on which to live their dream, and looked at 25 potential properties. The largest piece, at 27 acres (10.9 ha), appealed the most, even though its slope down the side of Mount Douglas made it the most impractical.

When they walked into Mrs. Smith's farmhouse on that Sunday in 1951, a little painting on the wall caught their eye. They read the familiar words printed beneath the image: "The Lord is my Shepherd. I shall not want." Widow Smith declared, "Well, I don't do business on a Sunday, but make me an offer."

Faced with an asking price of $16,000, the Chamberses made a cash offer of $15,000, which Mrs. Smith enthusiastically accepted.

A Farm's Evolution: The First Generation

Lawrence provided construction help to many of his neighbours, and with the money he earned from that labour, he and Ruth bought a herd of chestnut-brown cattle. Dual-purpose short-horns, they produced both milk and a thousand pounds (454 kg) of meat per year. One wet summer, Lawrence and Ruth had to turn the hay three times by hand, just to keep it fresh. With all the fresh fodder and adequate pasture, their cattle thrived.

After almost ten years of cattle farming, the Chamberses decided it was time to sell the cows. In 1963 the couple opened a vegetable stand. The product of Lawrence's green thumb meant that every afternoon from 1:00 to 9:00 pm, Greater Victoria customers flocked to buy a variety of peas, carrots, green and yellow beans, beets, broad beans, lettuces and, most popular of all, corn. Over the next ten years, the vegetable stand grew even busier, with a record sale one Labour Day of 1,200 cobs.

In the spirit of her parents, keen to protect both nature and farmland, Ruth drew on her Oxford University education in natural history and co-founded the Victoria Natural History Society in 1944.[50] She also wrote columns for Victoria's *Colonist* newspaper. As housing developments began spreading concrete over the best soil on both sides of the park, Ruth and her friend Irene Block helped the local farming families band together, establishing the Saanich Greenbelt Society in 1976.

As the geneticist Gregor Mendel observed when studying the genotypes and phenotypes in pea plants, traits often skip a generation. This also seems to be the case with farming. After spending almost every day of their childhood plucking rocks out of the soil, removing stumps and milking cows, Ruth and Lawrence's three sons had no desire to become farmers. Two became teachers who moved to the interior of BC, while the third worked as a civil servant in the Yukon.

As Ruth and Lawrence aged, the demands of increasing traffic, staff and wages became too heavy a burden. In 1972 the couple decided to scale back the intensive labour and grow hay. To increase the odds of success, they planted orchard grass, rye grass, clover and alfalfa. Other than some test plots at the island's agricultural offices, their seed from the Okanagan sprouted into the first alfalfa on the 20-mile (32-km) Saanich Peninsula. In 1982 Lawrence died, leaving Ruth in charge. After harvesting Madrona's best combined crop the next year in 1983, one of Ruth's grandsons worked on the farm with his granddad and their neighbour, Roy Hawes, for several years until he finished university. When he graduated, Ruth leased the farm out for hay production until 1999.

The Restorative Generation: David

Another grandson, David, grew up in the Yukon, snowboarding and working as a camp cook in his early adulthood. During a six-month stint walking, hiking and snowboarding around Japan, he picked up some gardening jobs, which inspired him to write a letter to Grannie Chambers about potentially trying his hand at farming. On his way home in 1999, he stopped in Victoria. Thrilled to connect with her, to share her knowledge and, best of all, to receive her enthusiastic approval to run the family farm, he returned to work up north, but only long enough to earn money for farm equipment.

Although hay had been continually planted and harvested for at least 14 years, the soil had received no replenishment, nourishment or rest and, as a result, had become deficient. In addition, invasive species such as Himalayan blackberry (*Rubus armeniacus*) and Scotch broom (*Cystisus scoparius*) encroached until even the access paths for the tractor had grown over. The farm seemed to yearn for someone who understood the importance of agroecology.

With no experience but plenty of open-minded determination, David spent the first year pushing back the overgrown brush. Next he erected an electric fence and installed what he called his "pig tractor." As David moved the fence, his pigs dug and fertilized 1.2 hectares (3 ac) at a time. Gradually, he expanded the fence and gave the pigs free rein up and down the farm's mountainous slopes. Once their tractor duties had been accomplished, he sold them.

Soon after his arrival, David visited his grannie's long-time neighbour, Roy Hawes, a former Saskatchewan farmer renowned for his twinkly back-slapping greeting and his positive attitude. Roy worked the land until he was 88.

"Whooeee, you're a working boy," Roy said at their first meeting, pumping David's hand while squeezing his bicep.[51] Roy knew how to bring people together, whether it was for hockey games on the Hawes's barn floor or gathering horseback riders to take his many horses up Mount Doug.

Working for his mentor and for other farmers in the area, David soon gleaned a wide variety of perspectives on farming. He learned the benefits of owning a really small tractor and a single-blade plow, and how to bargain for them when making the purchase. Roy kept him company as he bought his first tractor, and David still remembers driving it back down the Saanich Peninsula, hitching up his plow and proceeding to try to break up the front hectare. Before long, he had dug up his grandfather's irrigation pipes. In an almost archaeological surprise, he felt as though he were re-establishing contact with Lawrence.

David struggled with the tractor and plow for five days before Roy came over to give him some tips on how to plow more efficiently. The next year, other neighbours volunteered their more modern machinery; the job David thought would take about five days was finished in an hour and a half, thanks to that kindness. When they saw how keen David was, they were more than happy to lend advice, machinery and manpower to the young man whose grandfather had been their friend.

David soon realized it would take a number of years to

restore the farm to its former health. He also saw that the land all around him was no longer truly arable. After experiencing Roy's neighbourly assistance, David understood the benefits of educating the community about their food sources. As a start, David and Ruth contacted the local mental health services and invited their clients to refresh themselves by joining him in the outdoors, learning how to farm the fields and helping sell the produce they worked so hard to grow. Stretching their muscles in nature as they followed seeds from soil preparation to edible products proved to be beneficial for all of those who took David up on his offer.

David worked long hours every day. When a concerned grannie intermittently chastised him for overworking, he would declare, "It doesn't seem like work if you love it." On January 11, 2002, knowing that she had done everything she could to help support David through the "rockiest years" of his young farming life, Ruth Chambers died peacefully with David holding her hand in the farmhouse she had called home for 50 years.

The Restorative Generation: Nathalie

While David was bringing his grandparents' farm back to life, Nathalie's yearnings had sent her off in search of wild places, where she could practise conservation and ecological restoration. For ten years, she held a variety of jobs, including herbalist, tree planter, health-food store employee and volunteer assistant naturalist at the Swan Lake Nature House, where she took children on interpretative plant walks.

Although Nathalie was born in Winnipeg, she spent her first decade in Edmonton. Special trips farther north meant she could play at the Calling Lake cabin of her Uncle Wilfred. She still remembers every little tree that used to provide shelter and climbing opportunities. She still smells the yarrow and hears the hum of the big, blue dragonflies dancing around the bumblebees that were busy pollinating. She still treasures a vivid memory of herself lying on the grass, staring at the tall stalks of fuzzy, yellow yarrow as bumblebees buzzed on to their soft landing pads while lanky dragonflies circled overhead.

In Nathalie's family, farming also skipped a generation. At age 6, Nathalie's dad predicted that when he grew up, he'd leave the farm and fly military jets. He joined the Air Cadets as a young teenager and by age 17, he had begun piloting large jets in Germany. Nathalie's dad moved her family to Victoria in 1980, where he flew helicopters from the ship *Provider*.

To Nathalie's 10-year-old psyche, Victoria symbolized "a great wilderness of beauty, beauty, beauty." Even the names of places in the neighbourhood seemed to shout out how she felt about her new home: "Biodiversity Central." Their new home seemed like a dream house, located on a street near Lily Avenue, next to a field full of lilies where she'd sit by a great oak tree. Looking up at the sky, she felt soaked in a premature conservation ethic: she seemed to know innately how crucial it was for everyone to connect with nature. Often she'd run across the field, climb up the hill by Rogers Farm to the top of Christmas Hill. As she stood on the peak, munching freshly picked, wild blackberries, she loved surveying the wildness of Mounts Tolmie and Douglas, the Salish Sea and, on clear days, the surrounding islands framed by the coastal mountains of both Washington State and the BC mainland.

Gradually, however, she saw her wild, open environment start to fill up. First the field behind the church grew houses. Next, Rogers Farm grew concrete instead of vegetables. The trees skirting Mount Tolmie fell over with development. As more and more roads filled up with more and more houses, she began to understand a new maxim. Every time she experienced joy in nature, it seemed destined to be ransacked. She felt herself sinking into deep global grief.

Nathalie journeyed into adulthood when she took a job as a tree planter. As she felt the rhythm of her shovel digging into the scarred, barren hillsides, her hands rooting each tree into another patch of welcoming soil, her lips began repeating a mantra that seemed plucked from deep within her soul: "Mother Nature and I are one, bonded by the morning sun. Together we walk down green roads, turn the corner and our hearts explode. No vegetation can be seen; the land left in such a horrifying scene. This is where our paths depart, as it is here our duties start. Out of my bags, I pull my shovel, in an effort to fill the puzzle. I am blessed by the Mother."

By this time, Nathalie understood that she needed to connect to the land to keep her own cultural identity alive through the memories of her own land origins. It made perfect sense to begin pursuing environmental studies and agroecology at the University of Victoria.

Studying agroecology at the University of Victoria deepened her understanding of why chemical fertilizers are so hazardous for plants, humans, animals, sea life and soil. She learned that diversion of waterways removes natural homes for wildlife – often birds and insects that offer pollination services – and eliminates the source of enriching minerals normally transported by rivers and streams to soils. She learned that the use of tractors and other heavy tillage equipment only goes so far in readying the ground for new seeds. In the short term, the plow does provide space in the earth for the seed, but in the long term, too much plowing can have potentially irreparable consequences, such as soil compaction, which causes reduced aeration, the inability of plants to take up soil minerals, stunted growth and significant yield losses. She began to understand why and how communities – at all levels, in all species – most productively support each other in the long term. The more she studied agroecology, the more she realized that the sustainable solutions nature provides for farming endure, while industrial agriculture's approach only addresses isolated problems, providing "quick fixes" that inevitably create escalating side issues.

Falling in Love

While Nathalie spent years supporting forests and trees, David emerged from the Yukon at the turn of the century with an entirely different sort of appreciation of nature. At that point, it seemed hard for him – coming from a land of pristine beauty – to believe that climate change existed in the Land of Plenty.

One day, Nathalie took her kindergarten-aged son Sage for a children's tractor-ride event at the end of Blenkinsop Valley. David, the driver, greeted them with, "Howdy! Welcome to the Tractor Ride!" Mocking the farmer role as he swung an imaginary lasso, he said, "Alrighty, kids! Ya wanna go 'round agin?" Nathalie thought he was actually an actor, as he "Ma'amed" her and helped her down from the tractor. Surely farmers didn't *really* talk that way, did they? She went home and carried on with her studies.

Six months later, she happened to be shopping near the Blenkinsop Valley and stopped in at Madrona's farmstand. "Boom!" said Nathalie to herself as she eyed David. "Hey," she said to him, "you're that [tractor] guy!"

"Oh, you're in school?" said David, as Nathalie talked about some of her university courses. "Well, I've got my shovel in the soil, and that's teaching me how things grow."

"Okay," said Nathalie, "what about Thoreau and Muir, and

Fukuoka? What about the relationship between native diversity and the crops you plant? What about First Nations land-management techniques? What about restoration?" This first discussion ended productively. Nathalie pointed out that we can't really remember our roots – who we are or where we come from – when we disconnect from the land even for a generation.

"Did you know," she said to David, "that we had an excess of food here on Vancouver Island a hundred years ago, that we were shipping it off-island, but now with our current amount of farmland, we have only three days of food available at any given moment on this whole island?"[52]

Nathalie thought a day in heaven was botanizing and discussing wild edibles with her teacher, Dr. Nancy Turner. But when David talked of the frogs singing, of picking cloudberries while watching caribou and climbing mountains in the Yukon, love blossomed. As Nathalie ranted on about plant diversity, David shook his head, laughed and said, "You're just nuts."

After one of their debates during her daily visits to the farm-stand, Nathalie returned home to find David's radishes sprouting in her compost. "That's when I knew it was love!" she said. While studying her textbooks, which buoyed up her conservation ethic, Nathalie decided she would convince her farmer sweetheart of climate change and the benefits of organic gardening by making him her pet sustainability project. Soon, he invited her to write her papers at the farmhouse.

After a while, David began telling Nathalie a deeper personal story, the story of his coming to look after his grannie, Ruth Chambers. As Nathalie says, "any fellow who would look after his grandmother and run her farm – well, that had to be the stamp on the envelope." In a desperate attempt to share some of her university course material with David, she borrowed a book from the library called *The One-Straw Revolution,* one of several by and about her favourite Japanese farmer, Masanobu Fukuoka, a philosopher-scientist who studied plant pathology and worked as a customs officer inspecting plants.[53] During a sabbatical from his work in customs, the young Fukuoka pondered the human's relationship with nature. As he explains in his book, "When it is understood that one loses joy and happiness in the attempt to possess them, the essence of natural farming will be realized. The ultimate goal of farming is not the growing of crops, but the cultivation and perfection of human beings."[54] Resigning from his job, Fukuoka chose to work on his father's rice and mandarin orange farm, trusting he would find the best way to manage the eternal balancing act inherent in food production.

Fukuoka found joy as he worked on the farm. He came to revere nature's wisdom, and called his organic and sustainable methods "do-nothing farming." Yes, Fukuoka's process includes hard menial work, but overall he stresses trusting nature's millions of years of experience producing food and keeping the balance. Fukuoka turned more than 350 years of Japanese rice farming on its head by pointing out that extensive ponds of water are unnecessary if the farmer sows rice seeds in tiny balls of clay. He also suggested planting clover with the rice to prevent erosion. Next, he advocated using the rice plant's own natural byproduct – straw – as a mulch that returns the plant's own nutrients to the soil to provide nourishment for future growth. For further natural enrichment, he recommended planting barley as a winter crop. He cared for his oranges in parallel natural fashion, producing higher-quality fruit at a production rate comparable to that of his competition, without having to use their expensive machinery, fertilizers or pesticides. In his books, Fukuoka, who lived to age 95, entices his readers with his philosophy and history, his understanding of sustainability in nurturing both the land and human communities.[55]

Nathalie believed the ideas of this famous scientist and farmer would really appeal to David. Soon after she gave Fukuoka's book to David, they made a date for tea and enjoyed the Christmas parade of lighted boats in Victoria's downtown harbour together. He crafted one of his gourmet dinners for her. When he went to the Yukon for Christmas, Nathalie thought about him every day.

After his break, David took Nathalie for a romantic little boat ride on his pond, where he mentioned that his best friend (who happened to be female and beautiful in every way) was moving in. He also admitted he hadn't read the book (which was now overdue.) Nathalie said to herself, "Okay, you're being ridiculous Nathalie."

Even though David assured her that romance didn't enter into the roommate equation, Nathalie couldn't be convinced. "Get on with it," she scolded herself. "Focus on school." She stopped going to the Madrona farmstand

After a while, however, she found herself admitting, "I kind of miss him. We were having some really good conversations. Maybe it's good to just have a friend who is someone you can talk to." So, one day, she took herself back to the vegetable stand.

When David saw her, happiness beamed from every pore. Tentatively, she said, "Oh, I was just going out to my mom's house and I wanted to take her some cauliflower." But there was none, just as there had truly been nothing romantic about David's relationship with his best friend.

David showed her around the farm, held her hand, looked into her eyes and, with a remorseful smile, said, "Oh, I thought I'd really screwed things up." They kissed. Then he insisted that

Nathalie bring Sage over to ride his bicycle on the safe gravel driveway.

Building Madrona's Sustainability

As she observed David learning from his first four years of farming experience and switching to sustainable farming, they both grew passionate about the necessity to protect good land. "Protecting" meant preserving not only enough land to farm but the total natural habitat around it – including the trees – which nourishes the soil as well as the farmers. They firmly believed that residents of Vancouver Island could have a Vancouver Island diet if they practised sustainability instead of buying into the myriad expensive monoculture "fixes" that nature never really needed. They grew to understand that a local diet could be a reality everywhere, not just on the island.

The first time Nathalie stayed in the farmhouse, she felt "Grannied." She and David stayed in Grannie's old room, but Nathalie didn't sleep at all. She kept feeling Grannie's presence. Throughout the night, as Grannie's energy seemed to envelop her, Nathalie worked hard to convince her that she was a good person, that somehow she would steward Grannie's land.

Surprisingly quickly, Nathalie and David built their business up from about a hundred customers to over three thousand. Part of the restoration project on the farm included educating customers about the farm's latest sustainability projects, with notes on the farmstand's bulletin board and farm tours. Customers learned about the subtle but crucial interplay between pollination, soil micro-organisms and vibrant vegetation in Madrona's natural ecosystem. They learned that each bug in a fallen tree provides food for the birds living in the forest, and that frogs provide natural pest control, as long as they have clean water. Customers also learned that monitoring these singing aquatic friends helped the farmers assess Madrona's environmental health.

As David and Nathalie restored Madrona's air, water and soil, they also planted 450 trees in their new orchard. In 2006 they started planting a north-south corridor of trees, one leg at a time, to create a natural habitat for pollinators, prevent erosion, provide shade, act as a wind block and possibly provide a future source for wood agroforestry. Nathalie and David monitor the trees daily, and regular bird counts help them assess the success of the project. By the time Madrona became economically viable, the Chamberses understood that providing food to the community went hand in hand with educating – and being educated by – the community. As she worked at the farmstand, meeting all the neighbours, including old friends of Grannie's, Nathalie gained a sense of their importance and value. They opened their hearts to Nathalie, bringing her beautiful photos of themselves in younger years, images that Nathalie could completely relate to.

At Madrona Farm's fresh food stand, a place of community sharing, people took a stand. At the stand, in addition to everyday news, people talked about their health, their desire for good food and their passion to preserve land. Nathalie found herself falling in love with many of her customers, grieving with them, celebrating with them and sometimes mourning the loss of them. Farming had grown beyond the production of food to contributing to a community-encompassing gourmet dish. For Nathalie, it was the best job, and Madrona was the greatest place in the world.

Stars in Their Eyes . . . and Shattering Dreams

At the vegetable stand, the customers voiced memories of Victoria, Saanich and especially Gordon Head, [56] painting a picture of a time when farmland was abundant, farmland families flourished and so much food was produced on the island that it shipped produce to the United States. Times have changed. Now Vancouver Islanders receive the majority of their vegetables from the United States. Unfortunately, the southern tip of Vancouver Island, that perfect place for growing food – best climate, best farmland in Canada – no longer supports the farming it once did. Today, only crops of houses grow there.

Customer memories of lost farmstands and of strawberries glowing with sweetness gave David and Nathalie the inspiration to challenge the status quo and use Mount Douglas as an illustration of the divide between two very different land-use

Agricultural Capability

Organic soils used for agriculture are divided into seven classes.

Class 1   Capable of producing the widest range of crops because both soil and climate conditions are optimum; easy management.

Class 2   Capable of producing a wide range of crops; minor problems with soil quality and climate can reduce capability but don't pose major difficulties in management.

Class 3   Capable of producing a fairly wide range of crops if good management practices are in place.

Class 4   Capable of a restricted range of crops; soil and/or climate conditions call for special management of the soil.

Class 5   Capable of producing perennial forage crops and specially adapted crops; soil and/or climate conditions limit capability.

Class 6   Should be kept in its natural state as grazing land; cannot be cultivated because of soil and/or climate limitations.

Class 7   Incapable of soil-bound agriculture.

Source: "Agricultural Capability," BC Agricultural Land Commission, http://www.alc.gov.bc.ca/alr/Ag_Capability.htm.

concepts – and two different soil qualities. the soil at Madrona Farm, on the western base of Mount Douglas, shaped by three previous owners into its current, improved state, was only marginal in comparison to other farms on the eastern side of the mountain. The best (Class 1) soil – on that opposite side – was gradually paved over and became a neighbourhood of houses called Gordon Head. Madrona's Class 2 and 3 soils are a perfect example of how agroecological farming practices can amend soil. The Chamberses baby their soil after crops are grown and sold. Otherwise, fertility would leave the farm with its produce in the hands of the customers. Knowing it is crucial to maintaining good-quality soils, the Chamberses plant cover crops, and they mulch and nourish the spent soil with the proceeds from their mobile chicken house. They also regularly rotate their crops to reduce depletion of the same minerals in the same soil, and to amend new areas. When they began to think about how they practised farming, Nathalie and David realized that their farmstand might represent more than just a place to sell, educate and love. The stand carried the potential to usher in a turning point, a home base from which many concerned customers could spring into action to protect farmland.

Nathalie and David's forward thinking led them to the dream of protecting Madrona Farm forever. During the 145-year farming history of the area, attitudes had evolved from a reverence for Earth to a devotion to concrete. If they could protect Madrona Farm forever, the Chamberses felt they would be honouring customers and neighbours, living and dead, on a deep level. They shared their vision with lifetime residents and former Saanich Greenbelt members Gordon and Joan Alston-Stewart, who said, "You two have stars in your eyes."

Nathalie and David attended Saanich Council meetings, where they spoke about the importance of green buffer zones, areas protected from development that would help keep biodiversity high on farmland, which in turn allows such farmland to be sustainable and produce quality organic food. At the time, however, the price of farmland was swinging like an axe over every potential farmer's head, and some farm owners were anxious to offload their land to interested developers.

By attending and speaking at municipal council meetings, Nathalie and David put themselves in the line of fire opposite developers and families who wanted their old farmland released from BC's Agricultural Land Reserve (ALR). Meetings all too often progressed from informative discussions to screaming matches. After one meeting, they managed to just miss an angry right hook. At another, Nathalie received a death threat. Soon, they found themselves walking home looking over their shoulders, wondering how much it was *really* going to cost them to protect farmland.

In the spring of 2006, Nathalie completed a 60-page agroecology paper for a diploma in restoration of natural services at the University of Victoria. Including maps, an extensive picture gallery of Madrona Farm's growth, a restoration timetable and a short bibliography, her paper clearly explained the potential for bringing tired soil back into production through a positive relationship between farm biodiversity and environmental sustainability. Up until this point, council members and citizens had been saying, "If land isn't arable, you should be able to build on it."

She took her paper to the council meetings in an effort to help them all understand the importance of forests and other land in proximity to farms, and to explain how biodiversity buffers – birds, bugs, pollinators, water, carbon sequestering, soil micro-organisms – are essential to agriculture. She knew that without this understanding, citizens could easily be persuaded to give their permission to dump buffer land from the protected status of the ALR and set it up for development.

Prior to the 1973, establishment of BC's Agricultural Land Reserve, Madrona's front 1.6 hectares (4 ac) had been legally subdivided from the back 9.3 hectares (23 ac). Before Grannie died, she had given the total land title to her three sons. So, Nathalie sent her paper to David's father and uncles, who up until this point had allowed David and Nathalie to continue farming. The clock on their generosity, however, had begun ticking.

Sure enough, one of the uncles came down to the farm, waving Nathalie's paper in his hand. He burst into the little room that Nathalie used as her office. David, who was washing lunch dishes in the kitchen across the hall, could hear his uncle's loud voice. "As a mathematician," the uncle said to Nathalie, "I know this is absolutely not accurate!"

Nathalie looked stunned as he continued. "You don't own the land. So, you have no right to talk about sustainability. This is my farm, and I want to sell it."

As Nathalie heard dishes slipping out of David's hands and crashing to the floor, she thought, "Why is he talking to me and not David about this?" Out loud, she mustered enough courage to say, "Well, what about leaving a legacy for your grandchildren?"

"I'm going to leave my grandchildren money so they can go to school."

"What about the community?"

"Do you expect me to subsidize the public's green space by continuing to own this farm?"

"Wow," Nathalie thought. "He really didn't like my paper." Out loud, she said, "What about Grannie? She would have wanted it protected!"

"She made no stipulations in her will for that. It's all in your head!"

"I know she would have wanted it. She'd have been really proud to know that her farm would continue to feed the people on this peninsula forever. The healthiest way is to protect it all. You must know that the value of this land is beyond money for only a few kids' education."

"Well," he responded, "if you think you can find a way to keep all those values maintained – and still get my money – then go right ahead. You have one year."

"Above all, send the bees love. Every little thing wants to be loved." —SUE MONK KIDD[57]

TWO : Welcome the Bees

Caress the Bees

One morning when Robin stood in line chatting with other customers at Madrona's farmstand, several bees were buzzing around bouquets of black-oil sunflowers. As one zoomed past a customer's cheek, the woman dodged in fear. Robin decided to share one of her most thought-provoking experiences.

"I once rented a sunny suite in Gordon Head. The landlord's 3-year-old son, Malcolm, often popped in to visit. On this particular spring day, he picked up a bumblebee sitting on the rim of the bathtub. He placed it gently on his open palm. 'Let's go outside,' he mumbled and ambled into the kitchen. As he shuffled past me and my sink full of soapy dishes and headed outside, my mind raced, but my tongue flapped in a silent tangle of possibilities. I tiptoed behind him, water dribbling from my fingertips. Afraid to say anything for fear of causing the bee to sting Malcolm, I was also aware of a much bigger energy, a kind of indiscernible understanding that this little boy might have a lesson to teach me.

"Out in the sunshine, Malcolm sat down on the grass, cross-legged. He stroked the bee's back and chatted as though communing with an old friend. Long minutes later, the bee flew up and away, the boy gazing after it. Within a few more minutes, I was talking with his artist dad who said, 'Oh yes, Malcolm's always picking up insects. He's fascinated with them all, whether they're ladybugs or praying mantises, butterflies or bees. He's never been stung.'"

"Wow!" said the customer. "How could he not be stung?"

"Little Malcolm was a keen observer," said Nathalie, who had been listening to the story from behind the stand's counter. "Most bees don't sting. Malcolm and his parents' gentle, observant reaction to the bees shows a natural connection that the average city person has lost."

So, how did Malcolm get away without being stung, even when he was petting the bee?

Nathalie explained that male bees cannot sting. Bumblebees and lone bees are usually meek and only sting to defend themselves or their hives. Even if you do threaten them or their babies – by going too close to their nests or colony or food – they'll give you three warnings. First, they lift one leg. Then they lift their second leg. If you still aren't paying attention, they'll finally turn on their backs and show you their stinger.[58] Even swarms of honeybees sting only if they're pushed into it, say, if we selfish humans are stealing their honey.

Twenty Thousand Different Bee Species

Originally foreign to North America, honeybees were imported from Europe about 160 years ago. Although many people think all bees are the same, honeybees comprise really only a tiny number of Earth's hard-working pollinators. The world has twenty thousand distinct species of bees,[59] with North America accommodating four thousand different native bees and BC hosting about 450.

Nathalie and David welcome customers into their field, just to take the time out of their typically "busy as bees" schedules and observe. As W.H. Davies penned, it's crucial in life "to stand and stare,"[60] in this case to notice the differences between more than a hundred different types of bees hard at work on Madrona's 9.7 hectares (24 ac) alone. One can even follow Nathalie's example and lie down in the field, look at the sky, watch the birds and the bees, and listen to their songs, allowing their humming and buzzing to resonate.

"Bees make a marvellous study topic," Nathalie says with a huge smile on her face any time she's able to talk about them. "They bring to light the magnificence of ecosystems. Just picture that with every flowering of every native tree – wherever you live in the world – there is a native bee associated with it. Here on the southern tip of Vancouver Island, flowering begins with Indian plum, then alder, maple, Douglas fir, arbutus, and so on.

"In 2004, around the last time anyone sprayed for gypsy moths on Mount Douglas, we had a beekeeper here on Madrona with nine honeybee colonies. People were spraying under pressure from American purchasers who wanted to buy BC's softwood timber, but only if they could be assured that the wood was guaranteed free of gypsy moths. [Gypsy moths were imported to the US in 1869 as part of an experiment in enhancing silk production; they ended up defoliating and killing swaths of trees instead.] Although they claimed the spray was organic, the morning after the spray was applied, Madrona's beekeeper lost 99 per cent of his honeybee colonies. They were covered in a white mucus.

"From that day forward, I wanted to learn about bees, so we started using flowers to attract pollinators. We planted broad beans and sunflowers alongside the crops, and all of a sudden, a whole new universe of bees started flying in to Madrona Farm. Oh, my gosh! I didn't know there was more than one kind of bee. Some were enormous honeybees, almost the size of little hummingbirds.

"I started studying bees with help from The Xerces Society[61] and began learning about global pollinator decline and colony collapse disorder. In my quest to find out why bees worldwide were facing such disasters, I found out a lot about them. Soon, I started basically growing habitat for the bees. I stopped growing food for people and started growing food for bees. These were my years as a defunct farmer, lying in the comfrey patch listening to buzz pollination. You know, bumblebees and a few others have this amazing ability to pollinate by disengaging their wings so that their flight muscles shake their bodies at a frequency similar to that produced when you play middle C on the piano. Their vibrations cause pollen to fall loose from plants like tomatoes and peppers, cranberries and blueberries.[62] I just wanted to lie there, listening to their sonication. When I had a little more energy, I spent time taking photos of life through the bees' wings!

"Then I got a job with TLC [The Land Conservancy] as a native pollinator enhancement steward and worked with 90 farmers in British Columbia in their Conservation Partners Program (CPP). I did 15 pollinator restorations using a downloadable form at the Xerces website that helps you evaluate habitat."

The form asks:

Does this farm have all the needs? – Water? Hedgerows? Nesting sites?

Is it free of pesticides?

What are their practices? (Do they cut down 100% of everything, or do they leave a third standing for the pollinators to keep their resiliency?)[63]

The point of the assessment is to evaluate and identify potential pollinator habitat, the potential for increasing that habitat and, in some cases, restoring it.

At Madrona, Nathalie created a flowering cover-crop line that both amends the soil and feeds the bees, as bees cannot extract pollen from grains. Bee buffers are used for edges (such as field borders, strips within the farm produce or grass sections). She started using other cover-crop mixes so flowers would be available to bees during the early, mid- and late season. Spacing out the flowering times of those crops is of paramount importance, considering that bumblebees, for example, go into dormancy in November and are the first to come out in February. If they emerge and find nothing to eat, they starve. BC has 35 species of bumblebees.[64] The Chamberses make sure to feed their great Canadian bumblebee friends by providing a mix of early growing crops, for the bees come out even if there's snow on the ground.

To keep warm, 70 per cent of bees nest underground in sandy, silty, south-facing slopes, which often puts them in direct competition for shelter with humans, as is the case in the Victoria area, where almost every south-facing slope has a house on it. The remaining 30 per cent of bees nest in trees, stumps, snags or twigs. This information was extremely important when David and Nathalie were in council trying to protect trees on farms and convince people to provide biodiversity buffers in the agricultural areas.

After one of Nathalie's presentations, one of the local councillors championed her talk by writing "Farms need buffers" into the Official Community Plan. Bees became Nathalie's business, and she admits that she has had to grow in sophistication from her logging protest days to figure out how to convince farmers that having a rich bee habitat (i.e., having lots of native plants and trees and insisting on freedom from pesticides) will make them lots of money. She calls this "speaking the Love Language of Capitalism."

Nathalie gives presentations on bees to scientists, school children, gardeners, farmers and politicians. "Out of 40 of those presentations," she said, "only five people knew that honeybees were not native to BC. Ten people knew that there were 450 species of bees in British Columbia. Two people knew that native bees didn't make honey."

Nathalie loves to open her presentations describing pollination. "Basically," she said, "pollination is plant sex. Due to their stationary nature, plants have had to devise fairly creative ways to disperse pollen – the male sex cell – from flower to flower. Thus enters the big, dark, hairy bumblebee who is the intimate third party responsible for plant sex: transferring pollen from flower to flower."

Thirty-five per cent of the crops most prominent in the human diet are pollinated by insects, while 75 per cent of flowering plants require insect pollination to bear fruit or drop seed. Pollination is an ecosystem service essential to all, from birds to humans to grizzly bears. On Vancouver Island between 2009 and 2012, 90 per cent of honeybee colonies failed as a result of colony collapse disorder; the bees abandoned their hives and did not return. Because people take care of honeybees, we have accurate figures about their death rates and general livelihood. Native bees have no human keeper to document their lives. Scientists fear that native bees are suffering at least similar, if not worse, rates of decline. Pollinators are keystone species the world over. They are the canaries in the coalmine, signalling healthy or unhealthy ecosystems. The death rate correlates rather obviously with the increased and steady use of insecticides, as announced in the *Harvard Bulletin of Public Health*:[65]

64

"Sub-lethal exposure to neonicotinoids impaired honey bees' winterization before proceeding to colony collapse disorder."[66]

How do we bring back the bees? Nathalie replies, "Habitat. Habitat. Habitat. Create the habitat, and the bees will come. Bees are four times more attracted to native plants than others. However, while native plants are establishing, it is perfectly reasonable to give the bees what I like to call a TV dinner: nursery stock flower plants."

Flowers provide nectar – energy and water – to the bees and other insects that pollinate them. Nectar is often located deep within the flower, which protects it from being washed away in the rain or evaporating. While bees and other insects search for the nectar, they brush past the stamens and spread the pollen. To point the way to their treasure chests, flowers dress up in special colours and patterns that guide their pollinators to them – providing pollinators signage to their landing strips, just as is done for airplanes. Although we can often see some of these – the yellow stripes on the lower petals of a purple iris (*Iris hybrid*), for example – insects are able to see ultraviolet colours that are hidden from us. While we see the flower of a potentilla (*Potentilla norvegica*), for example, as solid yellow, insects see white petals with little dark-pink stripes pointing toward a flashy centre of pink-and-white polka dots.[67]

Why So Many Bees?

Pollinating many different types of plants in a variety of habitats demands a variety of pollinators. Madrona Farm, for example, has nine different types of soil. Different slopes with different elevations produce different vegetation that attracts diverse pollinators with the varying tastes and biological needs. Madrona is located in and buffered by the rain shadow of the Olympic, Cascade, Coastal and Vancouver Island mountains, so most of the rain has already dropped out of the clouds before the winds move over it. Madrona basks in a Mediterranean-type climate, which encourages the growth of distinctive trees like the Garry oak (*Quercus garryana*) and arbutus (also called the Pacific madrona) (*Arbutus menziesii*) that are so common here. Vancouver Island's southern peninsula attracts 137 endemic pollinators, mostly bees, which make it possible for farmers to grow a large variety of produce year-round.

Bees come in all sizes, shapes and states of hairiness. Their colours vary from brown to black, even to bright green or blue. Their coats are marked with a great variety of patterns, from spots to stripes to almost plain, bright orange, red, white, grey or yellow. Some bees are social, others anti-social. Some live in colonies; others are strict individualists. Some dig tunnels to make nests; others live inside plant stems.[68]

Some bees can tolerate colder temperatures, while others are attracted to plants that grow in warmer weather. Some have long tongues for the deeper blossoms, while others have shorter tongues for the shallower blossoms. Most bees see in ultraviolet, and they each vary in their attraction to different types of flowers. Some work many long hours, while others prefer shorter workdays. Some live for over a year, while others live for only a few short weeks. All these different types of bees match up with the many different shapes, sizes, scents and colours of flowers, trees, vegetables and fruit that need pollinating in different seasons.

As one example, researchers noticed that the orchard mason bee has a very short foraging range compared to the honeybee, but it can work longer and in colder temperatures. Over a five-day period in a cherry orchard, they observed orchard mason bees pollinating for 33 hours, while the honeybees foraged for only 15 hours. During cold-weather years, cherry trees serviced by the orchard mason bees produced lots of cherries, but orchards that had only honeybees produced no cherries at all.[69]

Compared to honeybees, bumblebees can increase the fruit set for tomatoes by 45 per cent and increase their weight by 200 per cent. The right species of bee helps trees produce bigger and more symmetrical apples, giving orchardists an annual boost of about $250 more per hectare.[70]

If an apple is kind of lopsided, it wasn't properly pollinated. Apples that have been properly pollinated have ten seeds. An apple blossom has five petals, like the bottom of the apple, which sits on five points. In the core of an apple, we see a five-pointed star, each having the potential to contain two seeds. That's ten seeds in total, if it's been completely pollinated by bees making multiple trips to bring pollen from another tree to each of the corresponding five petals.[71]

Can Fruit Grow without Pollination?

Surprisingly, apples can grow without thorough pollination, but they are often lopsided or not as tasty as those that have been properly pollinated. David calls Nathalie the International Bee Detective. If all ten seeds aren't there, she knows all the bees aren't either. Perhaps the nesting sites are too far apart, or someone nearby is using insecticides, so the bees aren't giving ample pollen or receiving enough nectar.

Each wild bee visiting a blueberry farm pollinates between 15 and 19 litres (3.4 and 4.3 dry US gal) of blueberries. Blueberry farmers call them "flying $50 bills."[72] However, in 1970, when the Canadian government substituted the colourless, odourless and almost tasteless insecticide DDT[73] with an insecticide called fenitrothion (closely related to wartime nerve gas) to kill off the spruce worm in a forest neighbouring some New Brunswick farmers' blueberry fields, 70 species of wild bees were wiped out. That year, a number of blueberry farms produced no blueberries. Average blueberry production in New Brunswick fell from 2,494,758 kilograms (5.5 million lb) in 1969 to 498,952 kilograms (1.1 million lb) in 1970 – a loss of 453,592 kilograms (1 million lb) per season. For 4 years, New Brunswick continued to lose 453, 592 kilograms (1 million lb) per season. Concurrently, neighbouring Nova Scotia's blueberry farm production, free of the insecticide, remained stable.[74]

What did the farmers do? They sued.[75] The forestry managers learned that they needed to take a more pollinator-friendly approach to sustaining both the forest and its dependent neighbours. It took several years for the effects of the sprays to completely disappear, but eventually most of the bees came back and the blueberry business began to thrive again. Instead of learning from this harsh lesson 30 years ago, we're still using insecticides and herbicides. As a result, we are continuing to lose numerous species of useful pollinators.[76]

We need to recognize that pollinators really do create a lot of income. Killing pests with pesticides kills the pollinators – and income that results from pollinated crops – at the same time. In 2013 a study from Cornell University found that pollinators, fertilizing just 58 crops, contributed $29-billion to US farm income in the year 2010 alone.[77] A study conducted by a non-profit group in the UK estimated the value of pollinators at around £430 million annually. This group is extremely worried about the bees dying off, not only for the sake of the bees or because they are concerned about general environmental decline, but also because they've already calculated that it would cost £1.8 billion yearly to pollinate everything in Great Britain by hand[78] – which farmers in China already do.

A comprehensive German study, at the Helmholtz Centre for Environmental Research, showed that the cost of pollination

68

services worldwide rose from US$200-billion to US$350-billion over the 16 years ending in 2009. Since 2001 the costs for crops that depend on insect pollination have risen much faster than the cost of the grass family of crops like wheat, rice and corn, which depend on the wind for pollination. Germany's Helmholz Centre for Environmental Research attributes three causes to the rise in price for these products: spraying crops with pesticides, applying an overabundance of fertilizers and destroying insects' natural homes by cutting down hedges and trees to transform those spaces into industrial-sized farm fields.[79]

When pesticide companies promote a product as a "simple fix," we have to stop and think about how complex it really is. The world needs all the different native bee species because each species pollinates one or more special type(s) of plant(s). Farmers need only 250 hard-working orchard mason bee females to pollinate a full acre of apples. But to accomplish the same work with honeybees, they require one or even two full honeybee colonies, each populated by tens of thousands of workers.[80]

Different bees have different abilities to adapt to temperature change. The 150 to 200 bumblebees in the average colony will all die off at the end of summer – except the queens, who hibernate in sheltered banks until spring. Of the 250 species of bumblebees in North America, a couple of dozen live in Canada.[81] Besides the bumblebees, a few other native bees actually overwinter in their original nests, too.

How Do Bees Survive Cold Winters?

Bees produce their own, natural antifreeze – glycerol – and to overwinter, they cover themselves with a thick layer of decomposed leaves or soil. Many line their nests with soft moss or animal hair. In Victoria bumblebee queens come out of dormancy in February, when the temperature averages between

3°C (37°F) and 8°C (46°F), although they prefer temperatures above 6°C (43°F.) Soon their offspring are busy helping farmers produce fruit and vegetables. But honeybees need a consistent 9°C (48°F) before they'll come out to work.

At rest, bumblebees' body temperatures match their surroundings. If it's cold lower down, they'll fly higher to get warm. They shiver to increase the temperature of their flight muscles. If you watch one carefully on a chilly day, you can see it vibrate – the muscles in its abdomen are pumping – as it tries to warm up. At 24°C (75°F), it takes them only a few seconds to get their temperature up to 30°C (86°F). But at 6°C (43°F), it takes 15 minutes of muscle pumping to warm up to 30°C (86°F).[82] Because bumblebees and other wild bees come out so early in the season, it's important to make sure they have something to feast on when they emerge. Some cover crops overlap the seasons, and in Madrona's environs – with such mild winters, access to good water and soil – farmers can plant and harvest crops year-round. Once a main crop is off the field in the springtime, they can plant cover crops like native sunflowers, bee balm, clover and chickweed. After harvest in the summer and early fall, they can plant cover crops like red clover, fall rye, hairy vetch, alfalfa, Austrian winter pea and/or fava beans. While helping provide habitat and food for pollinators, those plants also feed the soil with extra nitrogen. Another key cover crop is buckwheat, which can be sown in the summer or winter to add phosphorus and calcium to soil. Cover crops provide a third service: their roots and decomposing stems and leaves break up the compacted soil or heavier clay, which is common in Madrona's neighbourhood.[83]

Ideal Plantings for Bees

If we provide a wide variety of flowering plants for bees, they thrive. The wide variety of bee species is attracted to a large selection of food. Fifty to one hundred different plant species provide a gourmet medley: flowers of different sizes, shapes, colours, heights and blooming times. Merely ten of their favourite flowers – planted in large, easy-to-find clumps – can nourish our pollinators well enough.[84] In trade for pick-up and delivery of all the pollen, bees keep a portion of it, as well as the nectar. Some of them turn that nectar into enough honey to match their body weight every two hours.[85] Bumblebees save the pollen, a valuable source of protein, mostly for the larvae and for the queen, so she has enough protein to make eggs.[86]

If farmers insist on monoculture, once that particular crop has flowered, there's nothing else for the bees to eat. They either die off or – if they have the energy and don't have to go too far – they buzz elsewhere.

How Do Bees Quench Their Thirst?

Bees do receive liquid from nectar, but like us they also need a source of good-quality water. Madrona Farm purposely preserves a pond and stream for bees. Quite a few home gardeners provide ponds with rocks protruding out of the water so bees can crawl safely down to the edge of the water and drink without falling in.[87]

How Well Can Bees See?

Bees see a broader spectrum of light than we do, including ultraviolet, so they perceive colours that are invisible to us. They don't see reds – although bird pollinators do – but all the other colours of the rainbow attract them. Just at the height of pollination time, some flowers actually design and produce the "landing pads" visible only to those who can see ultraviolet radiation. To the bee, this welcome mat looks like a series of concentric circles or a series of coloured dots. The bee needs to stick its tongue into the central area of these circles to find the goodies.[88] We humans need special equipment to see through a bee's eyes, but at least we can smell the flowers when they're putting out a strong perfume. Perfume beckons pollination, attracting a plant's favourite insect or bird that also prefers it and that may fly at a specific time of the day or night. Madrona Farm has a honeysuckle with a nocturnal perfume so powerful that when David and Nathalie come home in the dark, they know it's beckoning one of the night pollinators.

How Far Can a Bee Fly Nonstop?

Flight range depends on a bee's size and the air temperature. Bumblebees are one of the most versatile species. As well as pollinating a wide variety of crops, they can fly farther, fly in cooler temperatures and fly in darker light conditions than many other bees.[89] Of course, they have to work harder when it's colder, so they don't tend to fly as far or as long.

However, some bees are so small that their flight distances are very limited. One of the smallest of the bee species is so tiny that it can fit on the eye of a bumblebee, 2.5 centimetres (1 in) long. Such diminutive bees are limited to flying 152 metres (500 ft) at a time. Bigger bees like bumblebees can fly more than 1.6 kilometres (1 mi), so they can manage the great expanses of monoculture fields. Monoculture has forced out many of the original pollinators, as they evolved in tandem with a variety of plants, allowing them to thrive on short foraging trips. Sadly, as a result, numerous species of wild bees are in decline.[90] The Xerces Society recommends an ideal of one hundred metres ("a few hundred feet") average distance between foraging and nesting sites.[91] At Madrona, the Chamberses use this guideline when planting the bees' preferred menu of veggies, fruit and wildflowers.

Carefree Native Homes

Besides locating their homes near their food, bees need a big enough habitat for a number of nests, not just one. About 70 per cent of native bees like to nest in the ground – not in the compost-rich soils of the garden but in sandy, silty soils that are easier to dig into. Leaving strips of untilled perimeters, or at least not digging too deeply, can be helpful. Like us, bees look for a warm, dry location to burrow into, but they need a house only the size of a big salad bowl.

On a larger scale, dry lake beds often provide opportunities for towns of nesting sites, as long as all-terrain vehicle enthusiasts don't suddenly decide to zoom out over them. Loss of nesting habitat and insecticides make up the two biggest threats to pollinators. We've destroyed a great deal of natural habitat with roadways, mines, factories, apartment buildings and subdivisions that have only grass lawns. If we want pollinators to help us grow our food, we must pay more attention to preserving our remaining fine farmland.[92]

The other 30 per cent of North American bees tunnel their homes into stumps or taller bits of broken-off trees. Some can chew into woody plant stems and twigs or soft pine trees. Bumblebees like hollow nesting sites, such as abandoned beetle tunnels (often visible in logs or in packed earth under the grass). They also make homes in deserted mice or rats' nests, or the undersides of piles of undisturbed leaves or grass clippings. Even empty birdhouses will do.

Artificial Bees' Nests Need House Cleaning!

People build all sorts of artificial nests for bees, from wooden boxes to mud blocks to holey mattresses, from plastic buckets filled with straws to wooden frames wrapped around bundles of hollow reeds or bamboo stems. These human-made bee homes must be kept clean or the bees face two major problems. First, fungi and spores grow in dirty nests, causing the bees to have stomach aches. Second, if nests aren't routinely cared for, pollen mites can sneak in, eat up all the bee larvae's food and starve them.

If we want to offer artificial environments, we need to understand that bees usually build new nests every year. So, if we are going to offer them the same nest, we must follow through with conscientious annual cleaning. If we offer straw-type nests, we can swap old reeds and stems for new ones. Or we can use straw that decomposes quickly, to naturally attract bees to a new home. When bees vacate nest blocks, we can submerge that type of nest in a bleach-water solution.[93]

Allow Nature to Lessen Our Workload

To lessen both farmers' and gardeners' workloads, it's best to keep – and even bolster –natural vegetation. If we plant and replant a variety of food and flowers from April right through November, we nourish the bees. Home gardeners should start any new garden areas with some mature plants – either transplanted or bought from nurseries – until their new seeds or younger transplants have matured and are well-established. When mowing, bee lovers leave one-third of their vegetation standing. Bees are thus welcomed to forage and pollinate, resulting in an artistry of flowers and gourmet produce.

If farmers and bees can work together to maintain as much of nature's assortment as possible, farms will be healthy and will endure. The more we all support each other, the richer we all become. Some people call it bio-cultural-diversity. Others simply call it biodiversity. Biodiversity helps us all return to our roots and support each other from the ground up. Dr. Vandana Shiva's 2012 address to the United Nations discusses the dangers of separating ourselves from nature. A quarter of a million Indian farmers have committed suicide – and more continue to do so – because they were hoodwinked into believing that planting genetically modified, patented seeds would somehow be more beneficial than planting nature's seeds, which have been plentiful for thousands of years. During her address to the UN, Dr. Shiva explained that farmers in India were offered free, patented seeds for a season or two, after which they were forced to pay for them without any option to freely harvest their own original seeds. Too late, they learned that they had given up their rights to sustain their small farms and could no long support their families. Dr. Shiva highlighted, as no one else before her had, the importance of small, organic, sustainable farms, as well as the rights of bees.[94]

To avoid plowing up good habitat for industrial farms, smothering it with concrete, or flipping it over to the barren juggling acts of investors, we must save the best farming habitats in their original, complex ecosystems. This will allow us to become food secure, to eat fresh, healthy meals, and to be independent from huge importing costs, which include unnecessary fuel bills as well as unfair wages in faraway countries. We certainly need to be much bolder in questioning the use of pesticides and genetically modified seeds, which, in addition to being costly, are routinely immersed in insecticides before being sold.

Understanding Insecticide Issues

The word "neonicotinoid" is formed from "neo," meaning "new"; "nicotine" from the oily, toxic liquid in the tobacco plant; and "oid," meaning "similar to." Bayer chose the name "neonicotinoids" for their family of insecticides that are chemically designed to be similar to nicotine but that affect the brains of those they target in a more toxic manner. Developed in the 1980s and introduced in the 1990s, neonicotinoids are sold under more than a couple of dozen names designed to attract buyers: Adage, Adjust, Admire, Arena, Assail, Belay, Clutch, Confidor, Flagship, Intruder, Merit, and Titan.[95]

Every year between 2009 and 2011, American farmers used almost 3.5 million pounds (1,587,573 kg) of neonicotinoids. In addition to being used on farms, neonics are being used in schoolyards, gardens, parks and city streets. Neonicotinoids can be applied as leaf sprays, seed coatings, soil drenchings or turf granules, or they can be injected straight into tree trunks. The most widely used group of insecticides in the world, they produce $1.9-billion worth of annual sales – forming about 25 per cent of the global agrochemical market.[96] Ironically, comparison studies of treated seeds versus untreated seeds of the same hybrid showed no consistent change in either pest problems or crop yield.[97]

In 2014, after several years of study, 30 internationally recognized scientists concluded their worldwide integrated assessment of neonicotinoids and fipronil (another insecticide.) Acknowledging the direct toxic effects not only on pests but on beneficial vertebrates, they pointed out that assessments neglect to look at the other, indirect effects of the chemicals.[98] The Task Force on Systemic Pesticides points out that neonics persist in the soil for months or years. As their compounds break down, new active ingredients can become more toxic with time, making them poisonous to beneficial species. In its conclusion, this task force stated that "some neonics are at least 5,000 to 10,000 times more toxic to bees than DDT. The effects of exposure to neonics range from instant and lethal to chronic. Even long term exposure at low (non-lethal) levels can be harmful." The task

76

force further explains that these nerve poisons can cause chronic damage by impairing memory, lowering fertility, causing flight problems, attacking the sense of smell, changing feeding behaviour, reducing food intake and causing organisms to become more prone to disease. Bees often forage less when they have been in contact with the poisons. Earthworms' tunnelling habits change.[99]

Typically, many consumers don't understand the long-term effects of the products they are using. A major example is the use of Roundup, an herbicide spray with a key ingredient called glyphosate that penetrates any plant's tissues, from the leaves down to the roots, killing its ability to manufacture amino acids that are essential to life. Designed by Monsanto in 1974, at first its major drawback was its incredible strength: at that point, very few plants were able to withstand it. Monsanto had to figure out how to genetically modify seeds so food plants and flowers could withstand it.

Monsanto's scientists discovered that a gene the geranium – a common garden flower – could provide crop resistance to Roundup. They isolated the gene, attached it to gold dust, blasted it through a fine mesh screen so that it penetrated the cells of the plants they were modifying, and the seeds of those plants became Roundup Ready (RR) seeds.[100] Asserting that it didn't want to lose its research investment, Monsanto patented these seeds and made sure that "crops produced from Roundup Ready seeds are sterile. Each year, farmers must purchase the most recent strain of seed from Monsanto. This means that farmers cannot reuse their best seed"[101] – as farmers have been doing for millennia – and that Monsanto can double its profit by selling not only the pesticide but also patented, sterile-plant-producing seeds.

By 2012, however, it became evident that plants had started

becoming resistant to Roundup. Some weeds, like the giant ragweed that Roundup had at first been so effective at combatting, had grown bigger and tougher,[102] through cross-pollination with some of the original RR plants.[103] As happens naturally, the process of natural gene selection produced a few RR plants capable of seed production, and those spread. If one reads the detailed history of legal actions involving the use of Monsanto's patented products, it's glaringly obvious that traditionally helpful farming communities have broken down.[104] As Hope Shand, the research director of Rural Advancement Foundation International, explained, "Our rural communities are being turned into corporate police states and farmers are being turned into criminals."[105]

In addition to a possible lack of understanding of such pesticides, many farmers either don't read or don't understand all the instructions in the products they buy. One of Monsanto's own studies, compiled from surveys with 1,700 farmers, showed that Canadian canola farmers are spraying Genuity® Roundup Ready® canola at rates 45 per cent above the recommended doses or outside the application window. As a result, they are losing three bushels per acre simply by assuming "more is better."[106]

Simon Fraser University scientist Mark Winston reported that "a typical honeybee colony contains residue from more than 120 pesticides."[107] At regular doses, manufacturers say insecticides such as the neonicotinoids are effective at killing aphids. But these poisons don't discriminate. Besides killing perceived pests, they attack natural helpers like ladybugs, bees, earthworms and birds. The effects of sprays on insects range between temporary paralysis and death. In Victoria, in the fall up until late November, bees go into dormancy – 70 per cent in the ground, 30 per cent in trees, including twigs, snags and stumps. Native bees have been known to make their homes in the twigs of blackberries and other bushes. Thus, when we spray berry bushes, bees die along with the ladybugs and aphids.

Even if ladybugs survive a pesticide onslaught, one of their prime sources of food – aphids – is gone, and they have to endure other effects of the spray. Typically, neonicotinoids permanently paralyze ladybugs. If we were to go into a field that had recently been treated with such an insecticide – and stand and stare – we would see ladybugs trembling, unable to walk properly.[108]

Between the 1990s and 2005 in most parts of North America, average bee losses ranged from 17 to 20 per cent (due to mites, diseases and other stress).[109] However, since 2006, annual mortality rates for bees have been ranging from 30 to 90 per cent.[110] Most research points to the use of insecticides and GMOs. A study on honeybees living in hives that are moved around the country to pollinate a multitude of industrial farms revealed high mortality rates.[111] These bees are dying from three causes: stress resulting from constant transportation over vast distances, working in and around neonicotinoid-infected blossoms, and a diet of high-fructose corn syrup (grown from GM corn, whose seeds were immersed in insecticides prior to planting).[112]

In early 2013, one of the prime manufacturers of GMOs and insecticides, Bayer, funded a year-long, $950,000 study in Ontario by two professors, one at the University of Guelph and the other at Dalhousie University, and concluded there was no connection between neonicotinoids and bee deaths.[113] A year and a half later, the British Ecological Society published a paper stating the connection between pesticides and bumblebee foraging behaviour.[114] Two days after that publication, a news announcement stated that Ontario was going to restrict the use of neonicotinoids.[115] With results being reported all over the scale, and keeping in mind our belated discovery of the dangers of DDT, it would seem to make sense to err on the side of caution with neonics.

Those who can't – or don't want to – wade through all the research may at least wish to take a look at Harvard's Micro-robotics Lab. Its researchers have been designing robotic bees to pollinate genetically engineered plants.[116]

And for those under-impressed by either the prospect of hand pollinating or having to pay for a bee robot to try to do a better job than nature's pollinators, consider the results of bee specialist Mark Winston's study:

> Recently, my laboratory at Simon Fraser University conducted a study on farms that produce canola oil that illustrated the profound value of wild bees. We discovered that crop yields, and thus profits, are maximized if considerable acreages of cropland are left uncultivated to support wild pollinators.
>
> A variety of wild plants means a healthier, more diverse bee population, which will then move to the planted fields next door in larger and more active numbers. Indeed, farmers who planted their entire field would earn about $27,000 in profit per farm, whereas those who left a third unplanted for bees to nest and forage in would earn $65,000 on a farm of similar size.[117]

To Understand, Stand and Stare

Meanwhile, the forested shoulders of PKols stand and stare. Hidden in tree branches and on tree trunks, in blackberry blossoms and on leaves, in the soil, in squash flowers and bean blooms, and around the edges of Madrona Farm's pond, a multitude of different bees buzz, their sturdy little legs thickening with pollen. Bumblebees everywhere are disengaging their wings from their flight muscles so they can make double use of those muscles to shake themselves more powerfully. Soon, their whole bodies vibrate with a sound similar to the tune of middle C – and they sonicate, allowing more pollen to flow.[118]

"She allowed herself the luxury of a good cry, figuring that her tears were mingling with the downpour to soak into the soil. It was relief. It was joy. It was the knowledge that she had overcome, and it spilled out with her tears onto the ground that she had toiled with, to become a part of the crop she had planted with her own hands. It had sought to defeat her, and she had prevailed. Now, she was permanently a part of it." —TRACY WINEGAR[119]

THREE : Taste the Connections

Fresh Is Sweet

If we could somehow shake ourselves loose from the constant turmoil created by endless research and counter-research, if we could disengage our muscles from those who threaten us with patent suits, we might find ourselves like the bees, free to vibrate with the sounds and sights of nature's gifts. As we experience the gift of being in the present moment, a buzz of tears might flow down our cheeks as the joy of real dirt drifts between our fingertips. What does it mean to be free to fly beyond a corporate mentality? What draws consumers to favour farms over grocery stores?

"It tastes so much better when picked fresh out of the ground," said one of Nathalie's customers. "I like the connection with the farmers – buying fresh and local seems more real. When I was young, I loved eating my mother's vegetables and having my own vegetable patch. So, perhaps this takes me back to that early satisfaction. But I can't grow all the wonderful veggies that farmers can, of course. So, knowing the source of my food and, especially, loving the flavours – these make all the difference."

"I was at a potluck the other evening," said another patron, "and saw a platter full of finger-size veggies. As soon as I saw the grocery-store-style of almost-identically round little carrots sitting amongst them, I got really frustrated. I knew these weren't real baby carrots. They taste so disappointingly bland. They were just cut down to look inviting. It seems like such a scam. I stared at them wondering, 'Why do people buy food based on appearance instead of taste?' We all know that these carrots, like so many other vegetables, have been flown thousands of miles and then driven in trucks to all the big supermarkets. By the time they get here, they lose almost all their taste."

At Madrona, all vegetables are picked fresh in the morning and sold within hours. At the end of the day, any unsold produce goes to the food bank.

But independent farmers do not have the freedom to sell their produce to numerous big institutions such as hospitals, mental health institutions or universities. Across Canada, three multinational food service corporations hold exclusive, multi-year contracts with members of many campus administrations. All campus food and beverages must be bought only from those corporations. As a result, students cannot set up independent food outlets, hold farmers' markets or invite more sustainable food into their cafeterias.[120] However, if students can catch the administrators of their campuses before the next food-service contract is signed, they may be able to write in a preference for local food.[121]

Big food corporations hold similar contracts with hospitals. For example, in Greater Victoria, the Vancouver Island Health Authority (VIHA) agreed to a $1-million signing bonus for strategic, five-year-renewable, food-service contracts with Morrison Health Care Food Services[122] to provide six thousand meals per day for patient and retail food services in their South Island facilities.[123] Morrison is a division of Compass Group Canada, which describes itself as "the country's leading food-service and support services company, with over 26, 000 associates working throughout the country and managed revenues of $1.8 billion in 2013. Employing 500,000 associates worldwide, our parent company, UK-based Compass Group PLC had revenues of £17.6 billion as at September 30, 2013."[124] As part of their contract, the Vancouver Island Health Authority does not permit locally grown, farm-fresh food to enter these facilities, yet they do admit that "Food satisfaction ratings . . . remain challenging and require significant investment of time by both the vendor and VIHA."[125]

When David tilled the land at Seven Oaks Tertiary Mental Health Facility, one of VIHA's facilities on Blenkinsop Road near his farm, he helped the patients prepare the soil in order to grow a beautiful vegetable garden. One of the patients at first thought his mental health worker was crazy for suggesting he try gardening, but after growing a 30-square-metre (323-sq-ft) plot of cucumbers, beans, radishes, lettuce and peas, this outdoor work-therapy program made him realize how much he loved growing fresh vegetables.[126] Due to the VIHA's exclusive contracts, these therapeutic gardening patients are not permitted to bring their food into the kitchen. However, they are allowed to graze in the field, take food to their own homes (if they are out-patients) or donate it to local food banks.

We are what we eat. How is a health-challenged patient's situation supposed to change if he or she is continually fed food that is nutritionally dead, full of GMOs, ridden with pesticides, picked green and ripened while travelling thousands of kilometres by air or truck? If local farmers were allowed to provide fresh food to patients, it would be a win-win situation. Groups of farmers could have a solid contract with a big organization, and hospitals would be able to provide the best food possible for patients. Imagine a world where our children, the elderly, ill and others were all fed beautiful, local, organic food grown by farmers known to everyone in the community.

A little research reveals that the average distance food travels to get to our mouths is over 2400 kilometres (1,500 mi).[127] David Suzuki has pointed out that industrial agriculture basically converts oil into food at the rate of six to eight calories of oil for each single calorie of food.[128] Holding up a bunch of Madrona Farm's carrots picked fresh that morning, yet another of Nathalie's customers volunteered this: "I just wish that the people who buy all the tasteless supermarket produce would at least give themselves the opportunity to taste your freshly picked, sweet carrots that are bursting with so much more flavour. And if it isn't carrot season, they should just be patient. Just seek out another, delicious, local vegetable that is in season – it'll give their bodies a rest from overdoing one sort and it'll cut down on import costs, never mind carbon emissions."

When they wait in line at Madrona's farmstand, customers notice a framed picture of a tractor filled with a huge diversity of vegetables. Beneath the image are the words: How far does your food travel? Nathalie says, "The customers who shop at our stand come here for a reason. Some are recovering from illnesses. Some, elderly, remember what good food really tastes like. Some are well educated. You can always tell the difference between a new customer and a regular customer. The new ones just come for carrots, while the regulars will take as much as possible of whatever they can within each season."

Much of southern Vancouver Island's good farmland has been lost to urban sprawl. Robin feels guilty about where she lives because she learned only recently that her neighbourhood contains some of the best soil in Victoria. Twenty-five years ago, she watched while investors converted a nearby potato farm to houses. A little while later, a series of tomato greenhouses were likewise consumed. Back then, she didn't understand that her community – and others around expanding cities everywhere – were slowly dismantling their positions as food-secure societies.

Gradually, witnessing the mounting losses, Robin began to understand that we can't ever retrieve those precious pieces of land. Her tipping point came when she saw some contractors throwing old refrigerators, stoves and even old hoses into a sweet little valley near her home. Bulldozers topped it up with boulders, rocks and dirt, so they could have some level land to sell. Even if we took the houses down, planting in that kind of wasteland could hardly provide the nutrients that took nature thousands of years to accumulate.

Some Victoria landowners are still lucky enough to have homes on healthy soil so they can plant vegetables and maybe some fruit trees among their flowerbeds. Five kilometres (3 mi) from downtown Victoria, dairy cows once munched fresh grasses, and now people still steer their cars off a street that retains its old name – North Dairy Road. Instead of turning toward a dairy barn, however, motorists swing into a vast, asphalt, parking lot to shop at Hillside Shopping Centre. The meltdown of farming history cries to be noticed in that road sign. Where did all those dairy cows go?

Is Industrial Agriculture Healthy?

The average North American eats 30.4 kilograms (67 lb) of beef per year. Because grain-fed, feedlot cattle are fatter than grass-fed ones, merely switching to pasture-raised cattle would reduce a person's annual calorie intake by more than 16,000 calories.[129] Eating grain-fed cattle also boosts the consumer's exposure to hormone-disrupting and cancer-causing chemicals, as cattle in feedlots may be injected with hormones twice weekly. Cattle fed grain increase the chances of passing along E. coli.

The E. coli problems are related to acidity in the bovine's first stomach. Micro-organisms in the first stomach ferment the food before cattle chew their cud. In grain-fed cattle, the fermented food is acidic, but in grass-fed cattle, it's neutral. A strain of E. coli has developed that can withstand the stomach's acidity. Acidity in the human stomach is also our main defence against undesirable bugs. But because this E. coli strain isn't bothered by acidity, it passes along to us when we eat any grain-fed beef that has been contaminated with it.[130]

As well, consider the connection grain-fed animals have to the monoculture methods employed by most conventional grain farmers. Industrial farms squeeze cows into prison-like feedlots, preventing them from roaming and freely fertilizing their natural pasture land of grasses. Vast, monoculture, grain farms continually are affected by increased soil erosion, decreased soil fertility and high fossil-fuel consumption. Each half-kilogram (1 lb) of grain-fed beef requires 4.5 to 7.3 kilograms (10–16 lb) of corn, soy and/or barley in the feedlot. If farming

86

practices on conventional grain farms – which divert 70 per cent of the world's available fresh water for such irrigation-intensive agriculture – continue unchecked, they will contribute to the extreme droughts now increasing with climate change.[131]

By contrast, cattle munching grass freely in pastures perform weeding both for themselves and for us. By nibbling off seed heads, they keep their preferred, tasty, native grass blades growing up toward the sun. Their manure, spread as they forage, returns nutrients to the field with no effort on the farmer's part.

Many of us also notice differences in the flavour and texture of free-roaming chickens and their eggs compared to battery-fed chickens and their eggs. Chickens left free to roam produce eggs with greater taste. Their yolks are a deeper yellow while their whites are thicker. Study two packages of chicken thighs or breasts at the grocery store and it's easy to discern the thinner, paler thighs of the battery-fed, caged poultry compared to the thicker, wider, darker-yellow meat of the free-range chickens. If we care about the lives of the chickens we eat, it's worth noting that farmers who raise chickens in confined quarters often have to de-beak them to prevent cannibalism.

Chickens are intelligent. In a natural environment, they use more than 30 distinct sounds to communicate with each other about their social requirements and class hierarchy. One study shows that chickens are able to anticipate the future 90 per cent of the time, as they make decisions requiring patience and anticipation for larger food rewards.[132] With total disregard for their innate intelligence or comfort, industrial chicken farmers stuff chickens into battery cages; they live out their lives in a space with walls and floor the size of a standard 21.6 by 27.9-centimetre (8.5 × 11-in) piece of paper. Never once in their lives can they stretch or flap their wings.

Contrary to any conceivable amount of common sense or caring in animal husbandry, several states in the US actually uphold "Ag-Gag" or anti-whistleblowing laws. These make it illegal to produce videos, take sound recordings or photograph inhumane treatment or conditions of animals and report about them. So far, Canada does not have any such laws on the books.[133]

Paving Over Our Food: Present and Future

As we dig into the history of Victoria's lost farmland, we see the impact both on farmers and on human health. Where we once grew blueberry bushes in healthy bogs, tomatoes on sunny hillsides or green beans alongside a salad patch waiting to be picked out of the rich, dark, naturally composted soil, thousands of customers now pluck cake mixes, instant pizzas and cereal boxes from grocery shelves. Subsidies to industrial farmers of commodities such as corn explain why, between 1985 and 2010, the price of high-fructose corn syrup drinks dropped 24 per cent (increasing the rate of type-2 diabetes in children as they consumed more), while over the same period the price of fresh fruits and vegetables rose 39 per cent.[134] As consumers buy cheaper, bland, prepared food revved up with salt and/or sugar – or beef fed on corn and other grains – they manage to convince themselves they can save time and money, without considering the long-term costs to their well-being.

In addition to these health concerns, with both fishers and farmers going out of business, the decrease in traditional agricultural practices leaves cultures on the brink of destruction. As mentioned earlier, over four million Canadians are food insecure. In Toronto today, every third child goes to school hungry. As most people turn to food banks only as a last resort (in fact fewer than 25 per cent of food-insecure households make use

88

of them), this means millions of Canadians are still truly down and out.[135]

In the four years between 2008 and 2012, food-bank use increased by 23 per cent in BC and 73 per cent in the Northwest Territories, with a 30 per cent average increase across Canada.[136] In the summer of 2013, BC's Minister for Children and Family Development claimed that child poverty had declined by 41 per cent over the previous decade.[137] If that claim had been true, the province would have come down to the lowest rate in more than three decades. But at the end of the decade between 2001 and 2012, according to the non-partisan *2012 Child Poverty Report Card*, which was using Statistics Canada's figures, 43 per cent more individuals in BC were using food banks.[138] BC has been crowned with the worst overall poverty rate of all Canadian provinces.[139]

Canada still remains one of the few developed countries without a national meal program for children. The United Nation's special food rapporteur, Olivier De Schutter, pointed out in May 2012 that "food banks are not a solution but a symptom of failing social safety nets."[140] Canada's Minister of Citizenship and Immigration Jason Kenney responded that De Schutter was wasting the UN's money by coming to a developed country with this information. But his comment only empha-sized the reason why the UN had sent De Schutter in the first place: "to help countries identify blind spots in public policies that would be easier to ignore."[141]

Mr. Kenney went on to say, "Canada sends billions of dol-lars of food aid to developing countries around the world where people are starving."[142] If we're apparently able to do this, couldn't we use at least some of those billions to keep matching Canada's minimum wage to our own rising costs of food and housing?

It appears not. Almost a year later, in the spring of 2013, De Schutter returned to Canada, trying again to give us a wake-up call. It seems our current government is intent on hiding the truth. The federal government scrapped both the 2009 long-form census and the National Council of Welfare.[143] Effectively, we put the blinders on our ability to collect, assess and compare the data for both farming income and social-assistance rates across Canada.

We need to redirect ourselves to a truly whole, sustainable approach, to vastly increasing our awareness of how we produce food. Ideally, we should focus on sustainable agriculture by stopping at roadside farmstands, enjoying the fresh air as well as the good food. By supporting those farmers who understand the importance of biodiversity, we reinforce the work of native bees, uphold the importance of native trees that provide habitat for local birds and insects that live in balance with each other, among many other factors. As we eat their naturally sweet, fresh food, we are acknowledging that trees are worth more standing upright than cut down. The overriding imperative is obvious: we must protect all farmland in perpetuity. For 40 years, BC protected 4.7 million hectares (11,613,953 ac) of land in the ALR, but in 2014, the provincial government planned to open a majority of that land up to mining.[144] BC still wears the crown for food insecurity. The price of farmland and the lack of support for farmers (only 1 per cent of our provincial budget goes toward agriculture) must be addressed.

Paying Trees to Keep Us Healthy

David Suzuki points out that, historically, we look at a temperate forest in terms of lumber and calculate its worth simply in those dollar values. But Suzuki looks at what that forest is "worth" alive and working. He reminds us how hard a temperate forest works. It captures, filters and supplies water. It helps prevent floods. It stores carbon. It purifies air. If we had to buy those services, we'd have to pay about $2,000 per hectare (2.5 ac) per year.[145] Farms located alongside forests experience additional benefits, including natural preservation of soil quality and natural homes for beneficial pollinators like birds, bats, a variety of bees and other insects.

Of course, all this labour should increase the value we place on a forest. If we add in everything, including the recreational benefits a forest provides, the trees could be receiving over $5,000 per hectare (2.5 ac) in wages every single year.

We need to treasure the biodiverse ecosystems all around us that naturally control excesses of one species over the other, instead of killing masses of life forms for assumed economic benefits – whether we kill trees for lumber or a few "pests" at the expense of many other beneficial organisms. Trees simultaneously provide shelter, preserve moisture and build soil as leaves fall and branches rot. We see this so obviously after an ecosystem loses its pollinators, so we must remember that food security – a nation or community's access to a sufficient quantity of affordable and nutritious food – depends on biodiversity on the farm and around it.

Keeping Farms Affordable

The high price of farmland remains the largest obstacle to food security on Vancouver Island, across Canada and throughout many parts of the world. Expensive farmland denies farmers the ability to produce food and thus creates food insecurity. Currently farmland located around cities is speculated at the same price as residential real estate, especially when negotiations may include the promise of a zoning change – making it enticing for speculators but completely unaffordable for young farmers.[146] As Jeff Rubin explained in an autumn 2013 article in *The Globe and Mail*, "The price of farmland in Canada has outpaced both residential and commercial real estate, gaining an average of 12 per cent over the last five years. In some hotspots, such as southwestern Ontario, the price-per-acre has been going up by as much as 50 per cent a year. Even pension plans and hedge funds have become players in the pursuit of prime agricultural land, interest that is only sending prices that much higher."[147] The pressures to rezone are being felt across the country, not just in southwestern Ontario and BC. Investors follow the old real-estate adage, "Don't wait to buy land, buy land and wait." They can pay $50,000 to $100,000 per 0.4 hectares (1 ac) for farmland that has potential for alternate use close to the city of Regina. While waiting for the land to be rezoned, they derive payback income for their mortgage by leasing the land out to farmers. Their patience is rewarded when the multi-family residential zoning goes through and their land skyrockets to $750,000 per 0.4 hectares (1 ac).[148]

As young farmers aren't replacing older ones, the multinational corporations are moving in, for the rising crop prices, the uncut timber and other natural resources on the land. Until we address this issue, we cannot hope to get the next generation out on the land, and our chances for food security will continue to diminish.

With such farmland price speculation near urban centres and potential mining centres, the pressures are mounting. In addition, changes to farming practices that have occurred over the last half-century – vast swaths of monoculture devoid of supportive habitat and overwhelmed by chemical pest and weed controls – have had a deadly impact on bees and their pollination success. All these weigh heavily on the ability of young farmers to take up the mantle.

Ideally, we should remove subsidies from toxic industrial agriculture, whose practices have devastated pollinator populations and escalated environmental and health-care costs. Also, as most of the produce from industrial farms is exported, and we import more, Canadian consumers are no longer taking in nutrients from sustainably grown, local food. Industrial farmers have to pay high royalties every year for "intellectual property rights" associated with the genetically altered seeds they plant. Nature knows better!

In the spring of 2012, after a 20-year period, the number of Canadian farmers under the age of 35 had plummeted from 77,000 to just over 24,000. We have to admit that industrial agriculture, complying with free-trade policies that encourage exports at local expense and top-down corporate control that pays the local farm the least, hasn't been the solution. Paul Slomp, the youth spokesperson for the National Farmers Union, said, "Parents who are farming are telling their kids it's not worth the stress and it's not worth the debt."[149]

So, what does the next generation of farmers look like? Despite the obstacles, Nathalie says, "Hordes of the next generation of farmers are educated. They practise sustainable agriculture: agroecology, biodynamics, permaculture and biological farming. They are eager to get their shovels in the soil. At Madrona Farm, we cannot keep up with the number of farmers who want to work here. The problem isn't finding people who want to farm. It's in getting access to farmland itself. We have a responsibility to the next generation and to ourselves to ensure that the farmland is available."

Actions We Can Taste

We need to let go of compulsive messages to buy, buy and buy without questioning our sources. People tend to comment,

"What do you mean there's a danger of food scarcity? My grocery store's shelves are fully stocked, from top to bottom! Aren't yours?" We'd love to see people reconnect in a nourishing way to the land, to a whole countryside of organisms that all participate to make food both tasty and healthy.

We can't keep tinkering with chemical sprays or chemical fertilizers that kill millions of microbes in the soil, destroy pollinators, slash the trees that offer shade, habitat and mulch, and affect the very water that we all drink. How can we remove benefactors from ecosystems and expect them to function well? Farming makes sense only when farmers champion biodiversity.

Great wisdom usually gurgles just beneath the surface when it comes to Earth matters. Whatever actions we take seem to snowball on us, either positively or negatively, so let's stop and think before we act. Take the example of hydraulic fracking. What actions could we have taken as soon as we learned that fracking poisons drinking water? Some said that bottled water was the answer. If each person consumes about eight glasses of water per day, the initial costs per person for bottled water amounts to $1,400 per year; if you drink tap water, you pay about 49 cents per year. Additional environmental costs are associated with the plastic bottles that carry bottled water. Our carbon footprint increases in both manufacturing and transporting the bottles. Landfills overflow. Chemicals from the plastic leach into the groundwater. Plastic trash islands grow in the oceans, killing birds and marine life.[150]

As multinational companies sell more bottled water, they are simply encouraged to purchase groundwater and distribution rights.[151] In the end, locals have no water rights, so they are unable to sustain themselves or their farms. Deciding to stop fracking and explore energy alternatives that guarantee the safety of both the quality and quantity of fresh water seems to be the wise direction to take. Changing the current manner in which conventional farmers deal with pests also involves taking a step back to re-evaluate the important factors.

The moment we become proactive, "they" becomes "we." We understand the suffering of those slogging it out on Bay Street, Wall Street or in the Beijing financial district. Our first and biggest obstacle is to collectively climb the mountain that is hiding what humans as a whole have lost as a result of practices like fracking and industrial farming. When a swath of big, old-growth trees is slashed, we're all reduced. We need to climb our densest peak, remove our blindfolds and take a look at the full view. From side to side, top to bottom, through the rocks, down into the oceans, we must recognize that we're all connected, and no one human, no one being, is any greater than the next.

Most of all, we need to reconnect our adult selves with the innate awareness we had as children. We were born with a joyful understanding of the vast multiplicity of our connectedness. Too often, however, we gave over this joy to those who would manipulate us into dividing our world into small, separate compartments, each with different, all-controlling recipes for managing life. If we thought otherwise, or still expressed the desire to explore other areas, it seemed to be an excuse for further segregation through mockery and bullying from peers and adults, which redirected us into a culture dominated by top-down leadership.

It became dangerous to question the authorities about topics like money, religion, politics, extreme sports, wars, peace missions, intellectual pursuits, mythologies, consumerism, classism or even such basics as how agriculture can best provide us with the food we eat. As a result, many of us grew up submissive, unquestioning, alone, obese and unhappy. None of us desire such passive isolation as adults. However, by severing our fundamental connection with the world around us, we've dug a hole in our soul.

We all need love, support and appreciation for each of our abilities and ethical intentions. If we all base our ethics on kindness and respect across the age, national, educational, political, physical, religious, business and racial spectrum, we can cultivate a natural diversity that can make our hearts and bodies vibrate.

As is so often the case, people are the problem – but they're also the solution.

"When we try to pick out anything by itself, we find it hitched to everything else in the Universe. One fancies a heart like our own must be beating in every crystal and cell, and we feel like stopping to speak to the plants and animals as friendly fellow mountaineers. Nature as a poet, an enthusiastic workingman, becomes more and more visible the farther and higher we go; for the mountains are fountains – beginning places, however related to sources beyond mortal ken." —JOHN MUIR[152]

FOUR : Embrace Ecosystems

Recognizing Our Broken Bonds

In 1935 Aldo Leopold and his family took a worn-out farm in Wisconsin and made it healthy again. Since he wrote *A Sand County Almanac* in 1949, his family's restoration story has sold over two million copies in ten languages. People from around the world now visit the Aldo Leopold Foundation to experience the land's vibrant, restored forests, wetlands and prairies. As he explains, "If the land mechanism as a whole is good, then every part is good, whether we understand it or not ... To keep every cog and wheel is the first precaution of intelligent tinkering."[153] Leopold's extensive writings and his family's ongoing dedication to conservation education inspired a truly effective revolution.[154]

How do we humans, the definitive problem-makers, connect the solutions to the problems that are hiding within ourselves? How big and fat do cancer, diabetes, heart disease, obesity and societal malaise need to grow before we make the decision to cease manipulating and tearing at our increasingly broken bonds with nature? Signs of ecocide are everywhere: oil spills suffocate marine life, chemical sprays kill beneficial pollinators and rising carbon levels contort our biosphere. When will we hear nature's screams? As government and corporations walk hand in hand in a seemingly robotic march, when will we demand policies that apply a consistent, environmentally aware ethic to stop the destruction of the place that has supported us so kindly for thousands of years?

Before we can take any action, we need to understand what we have done, and we must reconnect with nature. Every day, 160,000 fast-food restaurants serve 50 million people in the US, and 44 per cent of that nation's population go out for fast food weekly.[155] As the average person in the US now consumes 2,700 calories daily, the obesity rate has risen from 9.7 per cent in the 1950s to over 30 per cent today. Yet approximately 40 per cent of all food in the US is wasted by planting too much; culling food that is the wrong size, colour or weight; trimming too much; leaving it sitting too long or losing some due to malfunctions in the manufacturing plant; throwing out unsold produce in retail grocery stores; discarding excesses in restaurants; and immense trashing of uneaten food in households. The average US family tosses between $1,350 and $2,275 worth of food annually, yet only 3 per cent of it is composted, while the rest goes to landfills, producing 23 per cent of the country's methane gas emissions.[156] In 2008 in the US, "the medical costs for people who are obese were $1,429 higher than those of normal weight."[157] Putting these figures together, we can see that those 30 per cent of Americans who are obese are losing an average of at least $1,800 on discarded food plus $1,429 on medical extras, or over $3,200 annually because of a poor approach to food.

Despite living in the richest country in the world, Americans are the least healthy people among the world's developed nations.[158] Canadians spend $1,200 more annually on food and beverages than Americans.[159] But as Missouri Professor Emeritus of Agricultural Economics John Ikerd, explains, even though "Americans spend less than 10% of their disposable incomes on food, less than any other nation . . . the industrialization of agriculture has brought Americans cheap food but it has not ensured everyone access to good food."[160]

"We are paying a high price," explain the people at Food Addiction Research Education. "Industrialized farming has managed to dictate what foods are available to the public – with shocking results: the surplus of government-subsidized foods end up in our children's school lunch programs. Meanwhile, organic farmers struggle with the layers of bureaucracy and high costs to get their foods organically certified and transported to our grocers and tables. Americans currently pay three times as much of their income on healthcare as they did six decades ago."[161] They ask us to contemplate the provocative title of a paper written by two university professors: "The Perils of Ignoring History: Big Tobacco Played Dirty and Millions Died. How Similar Is Big Food?"[162] They also point out how big corporations call research "junk science," play down the role of good diet in favour of physical activity, state no need to change foods, and aim the cause of food addictions and obesity at the individual rather than the industry. In essence, we have been supporting the decapitation of the original ecosystems that once nourished us.

Government standards for tap water are stricter than they

are for the much more costly bottled water that the majority of Americans drink. Their discarded plastic bottles are heaping over land and sea.[163] It's past time to apologize to our ecosystems and educate ourselves on how our friends are helping to rebuild the future of farmland. Although the 25 per cent of North Americans who welcome community-supported farming projects may seem dauntingly few, farmers' markets have been growing in number by more than 16 per cent every year since 1995, showing an eager desire to improve the health of our children, our elders and our land.[164]

Thanks to scientist Rachel Carson's book, *Silent Spring*, we stopped spraying the once-popular insecticide DDT in 1972.[165] Thirty years earlier, when DDT application peaked in 1963 at an annual rate of 85.3 million kilograms (188 million lb),[166] geneticists already knew that some resistant mutants would gradually be selectively produced to replace the more vulnerable insects. Farmers, as David Suzuki explains, were already being set "onto a treadmill of requiring an endless string of different pesticides . . . Perhaps one or two insect species per thousand species are pests to human beings. Using a broad-spectrum insecticide to get at the one or two species that are a nuisance to humans seems analogous to killing everyone in a city to control crime – pretty crude and unacceptable."[167]

Despite the lesson learned from DDT, we keep repeating the experiment with other chemical assaults, thinking perhaps that the next time it will be "okay." A chemical spray endures for a long time, not just the interval in which it flies through the air and settles over the plants. After the plants die, the by-products of the spray are stored in the bodies of insects that have eaten the plants, then in the bodies of primary consumers that have eaten them, and finally in the bodies of larger predators (secondary and tertiary consumers) that have eaten the smaller predators. In the case of the predators, the chemicals become especially concentrated in fat, becoming more and more concentrated the higher up the food chain the chemicals travel. Scientists began to understand this process – now called biomagnification – in the 1950s and 60s, when eagles began to experience population decline.[168] They discovered that DDT was concentrated hundreds of thousands of times in the fatty tissues of animals at the top end of the food chain – whether in birds' glands or in women's breasts.

Although any unpicked plants (or their parts) decompose, the chemicals that coat them do not. In the end, the chemicals collect in the soil, concentrating over time and transferring from the soil to nearby forests, bogs, ponds, lakes, rivers and oceans, to the plankton and fish that live in those bodies of water, and up the food chain to the animals that eat them. In 1962 when brave Rachel Carson dared to publish her concerns in *Silent Spring*, DDT's manufacturer – Monsanto – and other chemical industry leaders viciously attacked her with smear campaigns, threats and attempts to stop the book from being printed, including injunctions, and much more.[169] Today, the same corporations continue to attack and silence any information that contradicts their biased reports on their products.

We can change this picture by choosing to uphold a different set of values. Rather than looking at ways to exploit, we can look at ways to honour remaining natural ecosystems, and wherever possible, restore damaged ecosystems. Yes, we have committed unthinkable acts of domination through colonization and wars against each other and against Earth – but despite this, every cell in every single body is crying to reunite. Inside ourselves we know that together we can create a whole that is greater than the sum of us all, that if we choose to look deep enough, we will find no need or desire to dominate at the expense of any one of our cells or at the expense of any one of us outside ourselves.

We know we each have the capacity to reconnect with nature. Taking a slow, deep breath, we can relax into the beauty spreading its arms around us from sunrise to sunset, from moonrise through the night. With a humble inner bow, we can gently, ever so respectfully, ask nature for help. Many fellow citizens of Earth have been sharing positive, nature-based solutions to difficult problems. We would like to offer a few examples of these to give you hope and propel you forward.

Sampling Bio-Friendly Solutions

The AskNature website offers an extensive source of bio-friendly solutions that come, of course, from nature.[170] AskNature shares videos showing how bees make ultra-smooth, gentle, flight landings. Watch how owls fly silently and prevent turbulence. Learn how slime moulds find efficient routes and create communication networks. Marvel in the actions of bull trout, which teach us how to conserve energy using vortices. Notice how the specialized blood vessels in a jack rabbit's large ears demonstrate efficient temperature regulation. AskNature allows you access to lessons that engineers have learned from water ferns, which retain a layer of air between each leaf's surface and the water; this knowledge helped the engineers to minimize drag and increase ships' fuel efficiency. Another video reveals that the natural prairie grasslands thrive because many different species co-exist, which helps each of them maximize resilience. It's smarter to share the load instead of gambling on one or two at the expense of many.

Nature is also waiting to give us free pharmacy lessons at the Animal Pharm, where you can find out why honeybees need more than just boxes on a monoculture field to survive.[171] A variety of plants in their neighbourhoods provide the resins that contain an assortment of bee-needed antimicrobial compounds. Bees line their nests with these resins to keep their families healthy.

In similar fashion, female monarch butterflies that find themselves infected with parasites will search for any one of the many species of toxic milkweed to deter their attackers. If a field's natural ecosystem has been left intact, the monarchs will protect their offspring by laying their eggs on the milkweed. As the monarch caterpillars emerge, they feed on these medicinal plants, which provide high levels of cardenolides, which are fine for the caterpillars but toxic to their predators.[172] Monarch caterpillars and butterflies are kind enough to forewarn predators of their bitterness via their bright orange and yellow colours. University of Michigan chemical ecologist Mark Hunter declares, "When I walk around outside, I think of the plants I see as a great, green pharmacy. But what also strikes me is how little we actually know about what that pharmacy has to offer. Studying organisms engaged in self-medication gives us a clue as to what compounds might be worth investigating for their potential as human medicines."[173]

Although industrial farmers consider butterflies to be second-class pollinators compared to bees, sustainable farmers view them as bio-indicators revealing the health of their farm's ecosystem. If we find monarchs in abundance, the area is vibrant. Planting a variety of flowering plants, such as clover (*Trifolium*), daisies (family Asteraceae: including asters [*Aster*], marigolds [*Tagetes*], purple coneflowers [*Echinacea*], and zinnias [*Zinnia*]), lantana (*Lantana*) and milkweed (*Asclepias*), enhances the beauty of the farm and welcomes a variety of pollinators and beneficial insects.[174]

Wolves have also taught us much about sustainability. As the short film *How Wolves Change Rivers* demonstrates, the animals' very existence changes soil composition and the flow of natural water systems.[175] George Monbiot, who narrated the film, is also a prolific author of environmental articles and books that encourage us to reconnect with nature, to think outside the box with wit and wisdom. A couple of examples of his work include the article, "How We Ended Up Paying Farmers to Flood Our Homes,"[176] and his book *Feral: Searching for Enchantment on the Frontiers of Rewilding.*[177] Monbiot helps us laugh at our past and current follies while taking us on a deeper journey to a future that may yet embrace the wealth of nature's heritage.

For all aspects of sustainability, we recommend the inspiring collection of videos, graphics, talks and articles by digital-media group Sustainable Man, which uploaded the wolf video recommended above.[178] The group points out how we can we can see through to the other side of stories that have kept us apart and integrate practical, sustainable values in the interconnected web of life by:

- using renewable energy;
- creating sustainable economics (old and new);
- utilizing permaculture to help those with small pieces of land learn to maximize their personal food production;
- encouraging urban organic farms to take advantage of nature's brilliant, whole, natural ingredients to grow a variety of remedies in contrast to industrial farmers' (or Big Pharma's) extracted and patented remedies;
- finding fun and creative ways to reduce, re-use and recycle;
- discovering cost-effective ways to live sustainably;
- sharing inspiring stories; and
- watching encouraging videos created by a variety of people – past and present from around the world.

As Sustainable Human embraces both global and local, sustainable diversity, we can go to this media group to learn from Indigenous wisdom, from ancient philosophers and presidents, from authors and scientists, cartoonists and comedians.

The 2014 Brazilian riot of 20,000 people crying out on behalf of millions of landless farmers reminds us that education is the key to action.[179] Each one of us must continue to educate ourself about where our food comes from and how we can best support our ecosystems so that we can continue to eat in good health. We encourage our readers to forage at the sites we've suggested in this section and to keep grazing in the vast ecosystem of informative, community-spirited publishing pastures.

In the late 1950s, without any money and trusting the kindness of strangers, Satish Kumar walked eight thousand miles (12,874 km) from India to Moscow, then onward to London, Paris and the US, delivering a packet of "peace tea" to each of the four nuclear power leaders. Settling eventually in the UK, he has since worked continuously as editor of *Resurgence & Ecologist* magazine, "promoting planetary well-being."[180] This magazine's authors discuss such farmland-friendly topics as rewilding, green living, grassroots action and the power of small.

"When human beings consider themselves to be the masters of the earth and have dominion over it," said Kumar, "they are more likely to abuse it and exploit it."[181] Thus, he founded Schumacher College at Dartington Hall Estate in South Devon, England, with goals for rural regeneration. Influenced by both Eastern and Western thinking, the college uses progressive forms of education, art and agriculture, and it welcomes students from around the world. Currently post graduate courses include these:

- Sustainable Horticulture and Food Production (growing low-input, diverse and resilient food systems);
- Holistic Science;

- Economics for Transition (achieving low-carbon, high well-being, resilient economies) and
- Ecological Design Thinking.[182]

Every student brings opinions and knowledge to share in co-operative learning, as ecosystems on our most sustainable farms prescribe. Thus, like Satish Kumar, one deed of highest impact is to make contact.

Making Contact

Kneel down. Touch the earth. The desire still exists, if you only give yourself permission to feel the soil beneath your feet and spend time breathing in a sunset with wonder and appreciation. It may take courage to re-engage with the child still hiding within you, especially with that yearning to connect to nature with uninterrupted joy. Try to position yourself low enough to actually lie down, on the grass or the clover or the moss. Pick up a little dirt between your fingers. Squeeze it slowly. Notice how wet, dry, hard, soft or crumbly it feels. How dark or light it looks. How strong or weak it smells. Breathe in the different perfumes of weeds, flowers, trees, bugs, birds and animals, and make wild guesses as to their origins. Study the art painted by the vast, capable sky. Listen for the buzz of the flies and the bugs, and especially the bees. Notice how everything reverberates through your body.

Are the frogs there? They are another "canary in the coalmine." Find them, for they tell you whether or not your habitat is still breathing happily. Lie still, listen for them, and when they start singing, feel their tone. When their comical vibrations start resonating inside your body, allow your voice to join theirs. Learn from the dance of interrelatedness how to truly waltz with another.

Welcome to your ecosystem.

Using Imagination

In a 1929 interview, musician and scientist Albert Einstein said, "I am enough of the artist to draw freely upon my imagination. Imagination is more important than knowledge. Knowledge is limited. Imagination encircles the world."[183] With imagination, we can make sustainable agriculture become the norm again.

Imagine, then, how thoroughly we're related. If we're brave, we can release our inhibitions enough to picture the geese and the deer as our older brothers and sisters, or our mentors. Then, as Indigenous peoples around the globe have been trying to tell us, what can these animals teach us? Well, geese illustrate the benefits of group support. Deer demonstrate the benefits of calm connection to a home territory.

By flying together in V formation, geese reach their goals 70 per cent faster than they would flying solo. They also fly farther together than they would alone. They see better when spread out in the V formation, and they experience smoother airflow. (Just as a boat's propeller turbulence can produce ocean waves, geese flying in a V create air waves that can make for a bumpy ride if a goose flaps out of formation.) By flying at just the correct angle, each goose receives a lift from the air flow of the beating wings ahead of her.[184] A female leads the flock when they take off for a flight.[185] When the leader grows tired, she relaxes and lets another goose take the lead. Geese honk both in greeting to each other and in moral support to encourage those in front to maintain their good speed. If one of the geese in formation becomes ill or hurt, two buddies drop out of the flock to accompany him down to the ground, where they stay until he recovers or dies, at which time the friends either join another group or find their original one.[186]

When facing possible foes, deer keep calm. They often take the time to assess an intruder, such as a human walking past them in the neighbourhood. They size up the individual before making a sweeping judgment, and they frequently keep on eating.[187] However, they eat only as much as they need, remaining svelte and healthy, lithe and graceful. In addition to being great swimmers,[188] they are able to sprint up to 48 kilometres per hour (30 mph), leap nine metres (30 ft) in a single horizontal bound or jump three metres (10 ft) straight up.[189] These skills allow them to act swiftly when confronting an enemy like a cougar or a hunter. Unlike the West Coast black-tailed deer, the midcontinent white-tailed deer notify others of danger by raising their tails like flags.[190] Deer are exceedingly adaptable, able to live in wild forestland, alongside farmland or in suburbia. Except in the Far North, where they may travel as many as 129 kilometres (80 mi) to avoid snow,[191] deer have no need to migrate because their home range varies from less than 1.6 square kilometres (1 sq mi)[192] up to 4.8 square kilometres (3 sq mi).[193]

In this quest to save farmland, we need to stay calm, to recognize that we have saved the first farm in perpetuity and that we can save many more. Right now, like some deer, we are raising these information "tales" to notify our readers of the imminent dangers of losing healthy soil and nourishing food. We want to encourage investors to support sustainability, to invest in the possibility for healthy food to be continually produced instead of buying farmland and holding it empty so land prices are forced to rise. We want to save the farms, wrap our communities warmly around them and invite children and adults alike to come and snuggle up close to the sources of their food.

In just one gram (0.04 oz) of soil, scientists have found a million species of bacteria networking to maintain life.[194] We must be careful about making assumptions as we deal with myriad other, often invisible, species and their complex interactions. Like the deer, our minds will have no need to migrate if we act swiftly to reconnect, to see and overcome any possible further loss by saving our sustainable farmland. When we acknowledge the valuable lessons we can learn from the deer, the geese and every other member of our biodiverse world, it makes good sense to continue to nourish and support sustainability. We are one, vibrant, complex whole.

Just as the geese must constantly adjust their V formation as they migrate, humans can realign themselves as they move along their separate or joined journeys. Perceived differences and grievances can be mitigated. We can share our skills. We can encourage one another while nurturing the sickened land back to health. Likewise, we can stay with those who have lost touch with our soil until they are healthy, too. We know that reconnecting disengaged minds will help everyone feel more energized.

Accommodating Biodiverse "Trespassers"

Even as we agree with the principles of biodiversity, the daily news keeps us aware of the problems of deer roaming in increasing numbers in our neighbourhoods and into farms, foraging through both vegetable and flower gardens. In the summer, swimmers at Beaver Lake near Victoria complain about the high levels of fecal coliform from geese. Geese and deer also move in and out of the Madrona Farm property, but the Chamberses are determined to remain inclusive. They work with the animals' natural behaviour patterns and set up special corridors for the deer, and special fields for the geese. Their cover crops attract a variety of both bees and geese. They direct the geese to those fields by covering other, fragile crops with blue twine to prevent them from landing. They also tie the twine to trees and stretch it across other areas that need protecting. After munching on cover crops, the geese defecate – and the poop from a hundred or more geese helps enrich the soil.

The Chamberses actually use the word "love" in describing the deer in their neighbourhood. Sometimes, the deer even sleep in Madrona's greenhouse. As Nathalie explains, "It's like a homeless shelter. Out on the two-hectare corridor, we feed them our leftover veggies. The bucks gather in bachelor parties and leave us gifts of their droppings to enrich our soil."

Deer pass knowledge along to one another, including about traditional trails. The Chamberses make sure to notice the locations of those trails and leave them accessible along a two-hectare (5-ac) deer corridor. Maintaining a good stock of the deer's preferred food along those routes means the deer will be more enticed to stay in those locations. At the same time, Nathalie and David protect their vegetable- and fruit-growing areas with fencing.

However, in the established neighbourhoods near Madrona, houses criss-cross the deer's traditional routes. It's common to see six deer cozied up on a neighbour's lawn. The local municipal district instructs homeowners to limit the height of their fences to a maximum of 1.5 metres (4.9 ft) in the front yard or 1.9 metres (6.2 ft) in the back, which deer can easily jump over. However, residents are allowed to have trees and hedging, as long as drivers on nearby streets have clear views, especially at intersections.[195]

Solutions could include installing open lattice work above a fence, upon which a homeowner could plant a combination of climbing vines, such as honeysuckle (*Lonicera*), clematis (*Clematis*) or climbing roses.[196] Such plants can quickly grow so wide and thick above a regulation-height fence that the deer can't leap up and over them. Another unusual solution involves interspersing a few taller pieces of bamboo between the regular 2.5 by 7.6-centimetre (1 × 3-in) vertical slats of the lower fencing material. It looks fine, can be spaced so it's easy to see through, and the deer won't consider trying to leap over what they perceive to be bamboo "spikes." Having a pet dog shouting a warning or two from behind the fence can also provide deer deterrence.

The Chamberses used the usual page-wire – that is, square galvanized wire – fencing, at least 3.4 metres (11 ft) high. They topped it with barbed wire and made a deer-proof wire gate to mark their no-go area. This created an understanding: they feed the deer behind the farmstand. When they leave, they close the stand's door so the deer can't take produce off the counter. The deer can roam the corridor but cannot pass the farm's gate. Sometimes the deer have created problems by breaking into prohibited areas and have had to be shooed out. But the Chamberses believe in allowing deer to be in the equation, so they don't shut them out of their natural habitat entirely.

Madrona Farm is relatively small. The Chamberses hear from farmers who have larger farms that it's impractical to fence or provide corridors. The Chamberses maintain that if a farmer can't manage the greater amount of work necessary to maintain his or her larger farm sustainably, he or she should share it or move to a smaller farm. One farmer claimed that deer were burrowing under his fences. Generally, only fawns squeeze underneath, as mature deer usually only dig down into snow to find food underneath. On rare occasions, they may dig in especially soft areas. If an extra 15 centimetres (6 in) of fence material is laid out flat on the ground, it can prevent animals from digging under the fence. If the fence is angled, or if a wide, angled hedge is planted, the deer can't judge the barrier's height. Some farmers have thoughtlessly buried or composted unsold produce in the ground right near the fence line, which of course tempts deer or other animals to dig for it. Any leftover food that a farmer doesn't give to food banks could instead be tossed to the deer on their side of the fence. In return, the deer will reward the farmer with poop.

As Nathalie and David say, "Poop every foot is good. This is brilliant for the soil, and in the case of the goose droppings, it revives a field for our use next year, without us having to bring manure in from anywhere else and without having to resort to fertilizers or pesticides. We've had our soil tested, and it's like gold! By making friends instead of enemies, we're all rewarded."[197]

Consulting Local Wisdom

Different human cultures have displayed significant genius. Polynesians, for example, learned to navigate the Pacific Ocean using only their multisensory observation powers and their brains, memorizing significant stars and learning to read the patterns of clouds and the waves and currents deflecting from distant islands. Americans put a man on the moon. Amazonians figured out how to live sustainably by the millions in the Amazon rainforest. Tibetans achieved enlightenment in only one lifetime.[198] Concluding a speech for *Seminars in Long Term Thinking*, author and anthropologist Wade Davis said, "The genius of culture is the ability to survive in impossible conditions."[199]

Internationally, tribal peoples have repeatedly proven to be outstanding custodians of their lands, having respect for biodiversity at every level. Native women have been traditional sowers of seeds and savers of land, understanding how forests, grasses and wild animals all work together to heighten genetic diversity.[200] Europeans would have done far better to consult and co-operate with Native peoples in the "new" land when they arrived in North America a few centuries ago, instead of ignoring millennia of local wisdom in their rush to colonize.

The colonists' extensive slash and burn farming practices violated the soil, upturned the helpful flora, killed the supportive fauna and started us on the road to increased carbon emissions. Admittedly, small-scale slash and burn as a method of agriculture has been used for about 12,000 years, beginning when humans changed from hunter-gatherers and settled in one area to farm their food. Today, about 7 per cent of the world's population uses this method, primarily for survival in areas of dense vegetation,[201] such as the tropical rainforests, whose warmth and moisture nurture 70 per cent of Earth's species.[202]

Trees in the equatorial latitudes of Africa, South America and Southeast Asia can regenerate quickly. However, the lifespan of some tropical trees is just a few decades, while others take more than a thousand years to reach old age. Sustainable farmers are aware of the need to honour the oldsters' genetic traits – fast rate of growth and pest-resistance properties, for example – by leaving them standing and gently clearing the land around them.[203] Otherwise, the resulting naked land permanently erodes: all traces of roots, natural ponds and streams, and their associated soil nutrients disappear. Worst-case scenarios have led to desertification.[204] If practised in the more northern or southern rainforest areas of the world, where the life cycles of trees range from two to seven hundred years, slash and burn can be devastating.[205]

When North American colonizers arrived, the ashes they produced from burning down the forests did provide some initial fertilization and free the soil of a few pesky weeds. But within a few years of farming the same plot with the same produce, fertility decreased while weeds again increased. At that point, the colonizers left such fields fallow, even allowing a short forest of bush to grow.[206] By the early 1900s, their progeny took their great-grandparents' approach one unfortunate step further and kept the deforested areas constantly clear, thus ensuring that their fields remained devoid of nourishment.

If they had consulted with the Indigenous peoples, they'd have learned far easier and more respectful methods. Keeping some of the trees, bushes, streams, ponds and beneficial habitat for pollinators, and spending a short period of time on each piece of land by rotating to others, would have given each area the necessary time to regenerate easily. To quote John Muir's wisdom passed on over a century ago: "Any fool can destroy trees. They cannot run away; and if they could, they would still be destroyed – chased and hunted down as long as fun or a dollar could be got out of their bark hides, branching horns, or magnificent bole backbones."[207] We can, and we must, open our minds to ancient wisdom to maximize our education and to ensure the preservation of farmland.

As Victoria ethnobotanist and Order of Canada recipient Nancy Turner explains:

> These people have, since time immemorial for them, adapted their lifestyles to the changing climates and the fluctuations in abundance of fish, wildlife and plants. If there is one overriding theme extending over time and space for human survival in this vast region, it is change – and adaptation to change. Being keen and vigilant observers, scientists in the broadest sense of the word, indigenous peoples have not only used the resources around them but maintained and enhanced them in various ways. They have developed ingenious and innovative methods for harvesting and processing their fish, shellfish, meat, berries and root vegetables for transportation and storage and have learned ways to optimize the nutritional and other benefits of these resources. At the same time they have found diverse means, through developing cultural institutions and protocols, to control their impacts on other species and to accommodate the changes that have occurred over millennia.[208]

Turner goes on to acknowledge that like all of us, Indigenous peoples have made mistakes in both resource and conflict management, but their oral traditions, which form a major part of their teachings, inform, warn and guide younger generations. In this sense, she speaks not only of biodiversity, but of bio-cultural-diversity.

As impossible conditions increasingly jeopardize our ability to grow truly healthy food, we can use our collective genius. We have the skills – and proven models – to help us find and save farmland in perpetuity.

> "Biodiversity starts in the distant past and it points toward the future." —FRANS LANTING[209]

FIVE : Encourage Biodiversity

The Gifts of Diversity

Nathalie realizes that with regard to food, starvation and poverty, every problem can be traced back to the massive loss of biodiversity worldwide. Even though food security issues differ and agriculture varies, it's the totality of biodiversity in each ecosystem that matters most. Starting out as a tree-hugging environmentalist gave her the overall vision to maintain her embrace of her ecosystem when it came time to dive into farming.

Ecosystems that are healthy include plants and their pollinators, trees that house insects and birds, and other organisms, all interacting in the whole, supportive neighbourhood. Honouring nature's biodiversity, as opposed to wiping out natural habitats with monoculture, ensures a farm's sustainability. When responding to Nova Scotia's agricultural land review, Pinnacle Farm's Rick Cheeseman said, "The only way to protect agricultural land is to farm it sustainability. Sustainability is one stop shopping."[210] Farmers who open their arms to this diversity are ensuring the long-term health of the land, maintaining its ability to raise a variety of food forever.

Sophie feels thankful for a present her parents gave her on her 11th birthday. Kick-starting her gardening career, they gave her a miniature greenhouse, a collection of pots they painted by hand, a watering can and packets of seeds. Early on, she learned the basic concept behind growing food: including a real, innate power, the ability to give life. As she tasted the satisfaction of eating the food she had grown, she integrated its consequence: self-sustainability.

If we want to nurture our children's ability to grow food and allow self-sustainability to thrive in future generations, we must take on roles as stewards. The remaining agricultural lands in British Columbia are the most highly valued lands because they contain watersheds, sensitive ecosystems and rare plant habitats – all of which contribute to creating nourishing soil that will grow healthy food. Thus, if we want to guarantee the greatest nourishment in the future, protecting farmland involves conserving whole ecosystems.

Before Europeans leapt into North America, Native peoples had been excellent stewards of the land for thousands of years because their cultures honoured the ecosystems.[211] In less than a couple of centuries, colonists have beaten these wise people up with guns, empty promises, sugar, alcohol and beliefs about how to govern. In this process of arrogant domination, assuming they had nothing to learn from Native peoples, the newcomers have also ignorantly beaten up the land.

Before Nathalie moved to Madrona Farm, she spent time with Betty Krawczyk[212] and other activists like Women in the Woods,[213] the Clayoquot Sound protestors,[214] Bear Mountain activists,[215] the Spencer Interchange protest[216] and many other political protests and activists. Her work on the logging roads prompted her to take up the activist's mantle. But she never felt she was progressing, standing on the logging road in the middle of nowhere with a sign saying "FU" to loggers. As Nathalie said, "It didn't really seem effective."

"While living in a tent on different clear-cuts as a tree planter," she said, "I started feeling sick. Gradually I realized that the whole camp felt the same. I couldn't really articulate it then, but I was beginning to realize that if we want to know how Earth is doing, we merely have to look at us humans. Or vice versa. We are one. Simply mirroring each other. Synergistically connected. And here we were, tree planters camped out on a hillside stripped bare of all its clothing, all its protection, all its own nourishment, all its little – and big – animal, bird and insect friends. Finally, it was obvious why I was feeling sick.

"I attributed the entire societal malaise, ecocide and all human-related problems, to the fact of our disconnection from the earth. All I ever got on the logging roads was chilblains. So, I decided I would use delicious-tasting vegetables to lure people back to the land. Surely tasting vegetables of that calibre would make people so curious that they'd just have to see how and where they were grown. Then I'd have them. It would be all bees and frogs from then on."

Recognizing the Problems

It seems easy for most people to recognize some of the problems created by logging and mining. The majority nowadays acknowledge that these activities must be planned and executed thoughtfully to support the environment. However, few people realize that industrial farming is similar to strip-mining.

First, industrial farmers strip off all the trees. Then they whip the poor soil to death with the same crops gorging on whatever beneficial nutrients they can glean year after exhausting year. As the soil becomes less and less nutritious, the monochrome plantings become weaker and weaker in their ability to be protected from their primary consumers, which in turn arrive in increasing numbers to dine on the more vulnerable

plants. Over and over again, as though drugged by the corporations who preach to them, industrial farmers reach for chemical fertilizers to bolster the soils and insecticides to kill the
primary consumers. The soils weaken while the insects
strengthen their immunities to the pesticides. In order to have
sustainable agriculture, lasting from generation to generation,
we must have biodiversity.

"Organic agriculture is the soup base of sustainability," says
farmer Rick Cheeseman.[217] In an appealing slideshow pre-
sentation, he pointed out to Nova Scotia's Agricultural Land
Review Committee that industrial food production contributes
20 per cent of the industrialized nations' entire carbon foot-
print. Using 70 per cent of all freshwater withdrawals, it far outstrips other industries' water usage (20 per cent) and domestic
usage (10 per cent).[218] When the International Assessment of
Agricultural Knowledge, Science and Technology for Development performed the first major global assessment of agriculture,
57 out of 60 countries supported it. The governments of Canada,
the US and Australia were the only three that didn't.[219] These
dissenters "accused the assessment of being 'unbalanced' and
attacked the authors' independence, despite the fact that all
dissenters were among the stakeholders who selected the
reports' authors in the first place."[220]

During his slideshow, Cheeseman pointed out that some
researchers at Cornell University studied corn and soybean
production for 22 years. Their results showed the same yields
for both organic and industrial farming methods. But organic
farming "used an average of 30 percent less fossil fuel energy,
conserved more water in the soil, induced less erosion, maintained soil quality, and conserved more biological resources."[221]

Cheeseman and the Cornell researchers are not alone in their
thinking. At the Manitoba Conservation District Association's

annual meeting in Brandon, David Suzuki made a similar point, saying, "Industrial agriculture is simply not sustainable. It is based on converting oil into food." He stated that pesticides are the "dumbest thing" that humans ever invented.[222]

With less big machinery, no sprays and therefore no spraying equipment, organic farmers save money. As the cost of oil climbs, so will the costs for industrial farmers who are mortgaged to their machinery. It seems ironic that organic food costs more than industrial crops. Government provides big subsidies to industry, while none are provided to organic farmers. Strangely, industry never receives fines or charges of any kind for the immense health and environmental damages it perpetrates.

If the consequences of industrial farming's short-sightedness were confined to one farm, that might be tolerable, as the farmer might learn from nearby mentors who could explain the economic benefits of methods of sustainability that include encouraging and maintaining biodiversity. But the monocrop mentality has spread over increasingly vast stretches of farmland. In Canada, between 1941 and 2011, industrialization ballooned the average size of a Canadian farm from 96 hectares (237 ac) to 315 hectares (778 ac) – an increase of over 300 per cent. During those same 70 years, the number of farms fell from 732,832 to 205,730 – a decrease of 72 per cent. By 2010, 5 per cent of the largest farms claimed 49 per cent of Canada's total farm revenue.[223]

This picture of a decreasing number of increasingly bigger corporations controlling more and more of the market has been repeating itself for decades. Consider the following facts. Three companies dominate the farm machinery sector,[224] so prices rise as choices are reduced. Fewer than ten companies control the chemical and seed sectors.[225] A "half dozen companies control . . . 75 percent of private agricultural research budgets, far outstripping any government's resources."[226]

It's important to note that "in the past 50 years, peasant agriculture has donated 2.1 million varieties of seven thousand crops to gene banks around the world. In the same time, seed companies have contributed just 80 thousand varieties."[227] As La Via Campesina's Guy Kastler[228] explains, "A plant is a living being. It adapts to where it grows, and peasants select them carefully according to their needs. Industrial seeds are selected to work in uniform conditions, they are not adapted to local realities; they're produced in laboratories and grown in test plots with chemical fertilisers." He concludes, "If we rely on corporate seed, we lose food sovereignty. If we lose food sovereignty, we lose political sovereignty."[229]

As a result of all this, farmers are facing increasingly reduced opportunities. Railways, grain-collection and distribution companies, beef and other packing plants, mills, breakfast cereal and pasta manufacturers, food and restaurant chains all increase their profits on the backs of the farmers whose own net incomes keep sinking.

Corporations make these glorified profits because, contrary to the farmers' situation, they do not have to factor in external costs. Some examples of these costs are:

- environmental damage;
- degraded nutritional values in the foods they produce;
- associated rises in health-care costs;
- increased water use and associated long-term losses (in terms of its future availability, its higher price and the consequences of having to deal with the poisons invading it);
- loss of both pollinator habitat and biodiversity;
- loss of traditional culture and language;
- increased poverty and despair among peasant farmers of the world; and
- weeds that are becoming resistant to broad-spectrum pesticides.

We can, and we must, choose to address this band of transnational corporations and the governments that favour them at our expense. The first step is educating ourselves so that we can make informed choices. One source of education is Pinnacle Farms, which portrays the status of the family farm clearly for all to grasp in an animated slideshow called "The Hamster Wheel: Farm Facts or Fiction?" As the folks at Pinnacle conclude, once we understand, we can "speak, write, vote and buy as if our lives depend upon it."[230] Because they do.

How Farming Has Changed in the Past Century

Over a century ago, while climbing with a shepherd and his huge flock in the Sierra Mountains, John Muir drew a clear image of the effects of large versus small sheep farming on the mental health of shepherds:

> The California sheep owner is in haste to get rich, and often does, now that pasturage costs nothing, while the climate is so favorable that no winter food supply, shelter-pens, or barns are required. Therefore large flocks may be kept at slight expense, and large profits realized, the money invested doubling, it is claimed, every other year. This quickly acquired wealth usually creates desire for more. Then indeed the wool is drawn close down over the poor fellow's eyes, dimming or shutting out almost everything worth seeing.
>
> As for the shepherd, his case is still worse, especially in winter when he lives alone in a cabin. For, though stimulated at times by hopes of one day owning a flock and getting rich like his boss, he at the same time is likely to be degraded by the life he leads, and seldom reaches the dignity or advantage – or disadvantage – of ownership. The degradation in his case has for cause one not far to seek. He is solitary most of the year, and solitude to most people seems hard to bear. He seldom has much good mental work or recreating in the way of books. Coming into his dingy hovel-cabin at night, stupidly weary, he finds nothing to balance and level his life with the universe. No, after his dull drag all day after

the sheep, he must get his supper; he is likely to slight this task and try to satisfy his hunger with whatever comes handy. Perhaps no bread is baked, then he just makes a few grimy flapjacks in his unwashed frying-pan, boils a handful of tea, and perhaps fries a few strips of rusty bacon. Usually there are dried peaches or apples in the cabin, but he hates to be bothered with the cooking of them, just swallows the bacon and flapjacks, and depends on the genial stupefaction of tobacco for the rest. Then to bed, often without removing the clothing worn during the day. Of course his health suffers, reacting on his mind; and seeing nobody for weeks or months, he finally becomes semi-insane or wholly so.

The shepherd in Scotland seldom thinks of being anything but a shepherd. He has probably descended from a race of shepherds and inherited a love and aptitude for the business almost as marked as that of his collie. He has but a small flock to look after, sees his family and neighbors, has time for reading in fine weather, and often carries books to the fields with which he may converse with kings. The oriental shepherd, we read, called his sheep by name; they knew his voice and followed him. The flocks must have been small and easily managed, allowing piping on the hills and ample leisure for reading and thinking. But whatever the blessings of sheep-culture in other times and countries, the California shepherd, as far as I've seen or heard, is never quite sane for any considerable time. Of all Nature's voices baa is about all he hears. Even the howls and ki-yis of coyotes might be blessings if well heard, but he hears them only through a blur of mutton and wool, and they do him no good.[231]

Over the past hundred years, the average farm size in the US has grown almost 200 per cent – from 60 hectares (147 ac) to 179 hectares (441 ac).[232] At the same time, farming style increasingly mimics factory production. To run any factory, machinery has become increasingly specialized, and these factory farms prove no exception. Although they need fewer horses and fewer people, they invest $97,000 for the average 160-horsepower tractor and up to $170,000 for the four-wheel-drive version. This explains why "Americans spend less on food than any other developed nation in the world – typically about 7% of their income – with only 2% of their disposable income on meat and poultry,"[233] a drop by half compared to their expenditure for the same products in 1970.[234]

To add specifics to our earlier reference on eating habits, it's notable that Americans also eat about 40.5 hectares (100 ac) of pizza per day, or 350 slices per second.[235] This kind of robust eating seems to parallel increases in farm size. Dairy farms may now number four thousand cows, while some cattle farms may host as many as a hundred thousand animals.[236] Over the past decade, American farmers added five thousand hogs daily to their factory farms, while chicken farms swelled by 5,800 every hour. The five states boasting the largest broiler chicken factories averaged more than two hundred thousand birds per farm.[237] The size of Canadian factory farms – and the treatment of animals within them – are just as startling to the average person who thinks of a farm as an innocent country enterprise managed by people who respect animal rights. "Over 95% of the 665 million animals raised and slaughtered for food in Canada today are mass-produced on factory farms. Here they live their short lives indoors in intensive confinement systems, deprived of everything that is natural to them including sunlight, family, and even the ability to turn around."[238] Those brave enough to watch the many photos and videos of factory farms online will receive a startling education. The number of factory farms is decreasing only because their sizes have been increasing – until now a significant few have gathered significant momentum.[239] The number of people involved in agriculture over the last 140 years has fallen from between 70 and 80 per cent of the US population to less than 2 per cent.[240]

Managing That Change, or Not

Of course, managing animal factory farms includes managing the disposal of high concentrations of manure, on farm property, on willing neighbours' farms or on other land.[241] As the US Government Accountability Office reported, ammonia, particulate matter and hydrogen sulphide discharge from these manure masses in "unsafe quantities."[242] Livestock and poultry industries generate half a billion tons of manure every year, more than three times more than that produced by the entire US human population.[243] In addition to the big burden on the soil, homeowners downwind from industrial hog operations have found their property valuations reduced by about 10 per cent.[244]

In Canada we have received bleak ratings not only for the extent of our animal cruelty on an industrial scale and the handling of manure but also for not meeting our obligation to provide information or follow-through on regulations. "Investigations conducted on behalf of the World Society for the Protection of Animals (WSPA) and our Handle with Care Coalition partners have demonstrated that it is not uncommon for animals (in Canada) to be forced to stand or lie in their own waste, in overcrowded conditions and endure extreme weather conditions without adequate protection, ventilation or bedding materials."[245] Environment Canada reported that, in 2001 witnessed the production of 177 million tonnes of farm animal manure was produced, much of which needed to be transported long distances to avoid health hazards (as was the case with the Walkerton, Ontario, tragedy in 2000, when seven people died and over two thousand became ill because cattle manure carrying E. coli 0157:H7 contaminated their drinking water supply. However, too often, industrial farms either leave it in large quantities locally, or illegally drop it in ditches or streams.[246]

"In a 'report card' released in April 2010, CFIA (the Canadian Food Inspection Agency) was given a 'D' by Canada's Information Commissioner, Suzanne Legault, for not meeting its obligations under the Access to Information Act 15 and awarded the

Code of Silence Award in 2008 from the Canadian Association of Journalists for 'its dizzying efforts to stop the public from learning details of fatal failures in food safety.'"[247] The Animal Protection Fund and Stephanie Brown of Canada's Animal Alliance both pointed out the extremely inhumane treatment of hundreds of millions of animals annually by Canadian agribusinesses and explained that Canadian authorities who are able to address the problem have done nothing to regulate or follow through with corrections and fines.[248]

Just as animal farms have supersized, so have crops. While dairy farmers milk four thousand head at a time, industrial farmers in the US plant monocrops of corn, soybeans, cotton or wheat, often over areas as broad as 4046.9 hectares (10,000 ac).[249] In 1996 about 15 per cent of farms in the US planted GM or herbicide-treated seeds. Fourteen years later, these figures had grown like weeds until a considerable majority of American farming acreage consisted of altered seeds: 60 per cent corn, 70 per cent cotton and 90 per cent soybeans.[250]

These same large-scale farmers may now manage many farms at once, not only around their home turf but overseas, as well.[251]

Why Did We Shift Gears?

The major shift from small-scale farming to industrial agriculture seems to have accompanied a mounting belief that factory-produced goods compete better in the marketplace. The mandate of the factory is to maximize productivity above all else. Newly industrialized farmers view monoculture in a calculated manner that prizes a certain economic model above all else. Their belief that front-end competitiveness should be isolated from – and superior to – long-term well-being allows them to ignore the health of biodiversity.

In order to compete in what they think is a successful manner in the market, industrial farmers concentrate on expanding only the highest-yielding, standard crops. They also process and promote their products in as many different ways as possible to repeatedly hook consumers at the supermarkets (and, lately, at alternative fuel stations some of which offer fuel derived from corn, vegetable oils and animal fats).[252] Such approaches mean turning a blind eye to: massive destruction of biodiversity and sustainability in farmland, obesity and associated malnutrition, the consequences of genetic components in food products that can adversely and irreversibly affect both human and environmental health, and the annihilation of diverse human communities that hold valuable agricultural knowledge.

We have clearly progressed to a point where "we cannot escape our predicament by simply continuing to apply methodological individualism, i.e., by relying on the outcome of individual choices to achieve sustainable and equitable collective outcomes."[253] The International Assessment of Agricultural Knowledge, Science and Technology for Development (IAASTD) points out that an "integrated approach to these urgent global problems" is needed. The group understands the necessity to include not only "agricultural knowledge, science and technology" (AKST) but also "economic and social science knowledge," including especially "the local and traditional knowledge that still inform most farming today." In the spirit of open-mindedness, IAASTD admits that even its AKST model "requires adaptation and revision. *Business as usual is not an option.*"[254]

When people query access to food and its growing scarcity in the same breath as industrial monoculture, Harold Steves Jr., a practising agricultural geneticist, explains, "During World War II people still knew how to grow their own food. We dug up backyards, lanes and roadsides. We grew 42 per cent of our food in

Victory Gardens. Everything was recycled. Manures and compost were returned to the land. Crops were rotated. Cow pasture was ploughed under to grow vegetables, then oats cover crop, then legume hay to add nitrogen, then pastures again. After over 50 years of monoculture, soils are depleted. It's scientifically impossible to produce more food with minerals and microbes that no longer exist in the soil. Yet that is the GM claim. GMOs do not produce more food to feed the world. GMOs are designed to kill weeds with Roundup or 2-4-D. That is all. To feed the world we need to maintain our genetic diversity, and replenish our soils."[255]

Is Incorporating Farms a Responsible Act?

Farms that have grown larger in size and in concentration of product per hectare, have tended to mimic the usual business model by incorporating.[256] The owners/shareholders of incorporated farms, as is the case with all other incorporated bodies, are no longer personally liable for the corporation's activities. "Unlike partnerships where partners are usually personally liable for the business acts of their partners, in a corporation, shareholders are not personally liable for the acts of the directors, officers or other shareholders (except in certain rare exceptions)."[257] With respect to small business in Canada, in particular, "You need to incorporate if your business involves potential liability that could seriously damage your personal finances. . . . As a sole proprietor, the [owner] could be wiped out financially, whereas the most a corporation can lose is the value of its assets."[258]

Similarly, in the US, a popular lawyers' website gives this explanation as the first in a list of the most powerful reasons to incorporate: "Protect yourself from personal liability." The site gives three examples. First, if a corporation signs a lease, "you're not personally liable." Second, if a corporation borrows money, "you're not personally liable." Third, if a corporation buys goods or services using credit, "you're not personally liable."[259] Thus, if corporations don't have to be responsible for the consequences of their actions, the responsibility is handed over to the individual consumers. We must each think seriously before buying food from corporations.

If we want farmers to be truly responsible and accountable, corporations as they currently exist should not be a part of the farming equation (or any ethical, caring business equation). Many consumers assume that all farmers truly care about the long-term effects on the health of their customers and the health of Earth. However, money does come into the picture. Once the corporations convinced some farmers that they needed to industrialize and ascribe to corporate business culture instead of nature's mandate of biodiversity and sustainability, the average farmer's net income slid disastrously. An average net income of $21,250 in the 1970s tumbled to $6,378 in the 1980s, slid to $4,516 in the 1990s and finally collapsed into the red, −$6,987 in the 2000s.[260] Although these figures are Canadian, the same story is true globally.[261]

Meanwhile, the biggest farming corporations, merging with their brethren so as to gain greater power, sharpened their profit pencils, distorted facts, captured supplies, destroyed competitors and enforced standardization. By thus suppressing competition, big farm corporations have been making record profits even as average farmers' net incomes have plummeted so low that they have had to take jobs in other sectors just to survive. Average farmers toil unaware of the fallacies in the stories they're being told, while the giant corporations slurp up all the profits. On the one side, profiteers coerce farmers into buying excessive energy, unnecessary fertilizers, genetically harmful seeds and harsh chemicals. On the other side, the packers, processors, retailers and restaurant chains slap their wallets over the farmers' heads.[262] Consumers need to re-educate themselves about the policies, actions and deceptions that occur all the way up the food chain and make more truly beneficial, local choices that honour the remaining farmers who truly care both about our health and that of Earth.

One specific example of corporate lack of responsibility stands out for us in the following quote implying that industrial farmers are making a profit: "The use of chemical fertilizers and pesticides has increased levels of farm output, lowered unit costs of farm production, and reduced pest-related waste and spoilage due to weeds and insects. Agricultural research leading to genetic improvements in animal breeds and crop varieties has also contributed to increased agricultural productivity."[263] Unfortunately, in the spirit of a general corporate lack of responsibility, the writers seem to stay conveniently silent regarding who should be paying for the consequences of the pollution, both in terms of consumer and ecosystem health, all the way down the chain of resulting loss and death.

In fact, the concentrated residues of both chemical fertilizers and insecticides have huge downstream effects. They wash into waterways close to industrial farms and then into connecting ponds, lakes and rivers. The fertilizers and insecticides pollute the water and soil of neighbours and neighbouring states and provinces. For example, the nitrogen compounds from farming in the Midwestern States are flowing all the way down the Mississippi River and creating a large "dead zone,"[264] killing off the fisheries in the Gulf of Mexico.[265]

A dead zone is an aquatic environment that has been robbed of its oxygen. Fish and other organisms mostly die of suffocation; they have either no time to flee the oxygen-deprived water – in the case of shellfish unable to move very fast or very far – or no way to migrate far enough away. The few organisms that somehow survive are usually sexually incapacitated by the time they get out of the zone. Dead zones contain concentrations of chemicals from a variety of sources, but mostly from synthetic fertilizers, typically those high in nitrogen, phosphorus and potassium.

By far the strongest poison in this chemical cocktail is nitrogen, which pours off chemically fertilized fields into streams, then into rivers and finally into oceans. When it reaches the ocean, that nitrogen provides the perfect habitat for a variety of algae to bloom in enormous bunches. Eventually, the algae sink to the bottom where bacteria start working at decomposing them. The bacteria need oxygen, and as they work on the algae in huge numbers, they rob the lower-level ocean water of its oxygen.

In the summer of 2013, scientists thought the Gulf of Mexico's dead zone might become the largest on record because of the heavy rains that caused a lot of flooding farther north.[266] Researchers logically expected more of the farm fertilizer to rush downstream. When unusually strong winds drove some of the water eastward, however, the southern dead zone was smaller than expected. However, it merely worsened the eastern dead zone, compromising the once-healthy Chesapeake Bay.

Since 2001 the South has been trying to reduce the dead zone to just under 3219 square kilometres (2,000 sq mi). Sadly, they've never even come close; the dead zone's area averages a little over 8047 square kilometres (5,000 sq mi) annually. The dead zone grew closer to 9656 square kilometres (6,000 mi^2) in 2014, three times larger than hoped and about the size of Connecticut.[267] No fish. No crabs. Not one living aquatic organism in the once densely populated stretch of ocean. Chesapeake Bay has suffered a similar fate.[268] Why should agricultural farmers have the right to kill fish – and in the process, kill business for fish farmers?

The problem is growing. In the 1960s scientists documented 49 dead zones.[269] In 2004 the United Nations Environment Programme reported 146 dead zones worldwide. By 2008 dead zones numbered 405.[270] In 2013 marine scientists reported 530 dead zones. In the Baltic Sea alone, fish have lost 1.3 million metric tonnes (1,279,468 tons) of food each year because of dead zones.[271]

In addition to the downstream, dead-zone factor, the practice of industrial agriculture is killing off local frogs and other beneficial aquatic wildlife, which would otherwise eat unwanted insects. As a result, in their rush to mass-produce tasteless grains and vegetables as cheaply as possible, industrial farmers simultaneously mass-produce more insects. So, in vicious circles, out come the chemical pesticides, again.

The insects that do survive the pesticide onslaught grow more and more resistant to the pesticides. Blindly, however, industrial farmers, their families (and perhaps their neighbours) may not question the quality of the air they breathe. Human bodies can't figure out how to resist pesticides. Residual pesticide buildup affects human hormones, reproductive abilities and endocrine systems.[272] Everyone eating GMOs is absorbing and reacting to the genetic alterations in the food. Through horizontal cell transfer, pesticides collect in our bodies, with some being excreted and lots being stored.[273] Side effects should be listed on the labels of food made with GMOs. These include increased chances of cancer, neurotoxicity, food allergies and autism.[274] Monsanto's lawn weed killer contains known carcinogens. Picture all the children playing and rolling around on those lawns.[275]

When people buy cheap food or practise home farming and gardening with pesticides, they are not thinking about the hidden costs. The Johns Hopkins Bloomberg School of

Public Health published an estimate of the environmental and health costs associated with the use of pesticides in the US: $12 billion per year.[276] If funding is tied to industry, however, some scientists (including those from high-profile universities), will "prove" and publish "results" that they then pass over to the media, claiming "scant evidence of health benefits from eating organic foods." We must all diligently research the broad field of evidence to weed out the spin from the pure research.

"Scant evidence" claims were made by Stanford University researchers in 2012.[277] The Cornucopia Institute stated:

> Academics and organic policy experts, including [those] at Cornucopia, immediately recognized that Stanford's research in fact substantiates dramatic health and safety advantages in consuming organic food, including an 81% reduction in exposure to toxic and carcinogenic agrichemicals . . . So we were not one bit surprised to find that the agribusiness giant Cargill, the world's largest agricultural business enterprise, and foundations like the Bill and Melinda Gates Foundation, which have deep ties to agricultural chemical and biotechnology corporations like Monsanto, have donated millions[278] to Stanford's Freeman Spogli Institute, where some of the scientists who published this study are affiliates and fellows.[279]

Tomatoland, a bestselling book by Barry Estabrook, an investigative journalist and a contributing editor to *Gourmet* magazine, clearly portrays how unkindness plays itself out in the nasty business of industrial agriculture.[280] Estabrook has travelled the fields of industrially grown, Florida tomatoes. He describes how bland tomatoes are grown in sand and fertilizers, sprayed with an assortment of pesticides, picked hard and green so they can be tossed in trucks and shipped to stores up north in time to ripen when the big-chain grocers finally place them on their shelves. Mexican immigrant workers spend 12 hours per day slaving under the spraying in the fields, returning home soaked. The sprays shorten workers' lives, and they affect pregnant women, who deliver babies with multiple birth defects.

Readers of *Tomatoland* come away fully understanding that we must mandate biodiversity if we want to stay healthy or attain health. The book exemplifies how fully humans have divorced themselves from nature. This severance is at the root of one of the world's biggest problems.

Time for Solutions

Our actions clearly cause a triple loss: of healthy farmland, healthy oceans and our own healthy bodies. How long will we be able to survive all the chemicals being pushed into our food, water and bodies? As the number of fish we have to eat declines, we face the horrible irony of water deserts.

Three easy solutions still exist to reverse dead zones. First, farmers can add at least a couple of different crops (such as alfalfa and wheat) to their corn and soy rotation. That will lower nitrogen needs by up to 80 per cent without lowering productivity. Second, they can plant cover crops from the grass, legume and brassica families after the fall harvest to prevent erosion and stop runoff so waterways are kept cleaner. Third, they can free their chickens, pigs and cattle. We must refuse to buy meat and animal products from farmers who insist on stuffing both small and big animals into cages. Confined, fed grains and hormones instead of a natural diet of grass (and bugs, in the case of chickens), factory-farmed animals' lives are a horror. As well, their chemical-filled dung is shovelled into one enormous, stinking mountain. When it rains on that poisonous pile, a lot of it runs downstream. Farmers must stop treating farm animals like factory products. Instead, they should allow them to trot around, freely grazing on natural food in sustainable pastures. While they're outside happily munching, they'll be distributing their much more natural manure over a wide area, building healthy soil.[281]

With a mind to resurrect the status of average farmers, in July 2005 the National Farmers Union of Canada submitted a report, "The Farm Crisis: Its Causes and Solutions," at a meeting of the Ministers of Agriculture in Kananaskis, Alberta. They recommended many sensible solutions. Foremost among them, farmers should be guaranteed their costs of production and land should be set aside for their farms. Of course, the power and profits gained by farm-supply manufacturers need to be controlled. Further suggestions include facilitating farmer-owned, co-op manufacturers; moderating suppliers so farmers can unhook from profit-drainers; banning corporate farming; instigating transportation legislation to restrain the pricing power of Canada's two railways; building costing reviews into their plans to help farmers succeed; capping revenues so profits cannot be inflated; and controlling excesses of retail power.[282]

One touching solution proposed by the union includes the necessity for educational labelling, requiring "that food labels disclose 'the farmers' share.' Toronto dentists, Halifax teachers and Vancouver parents, struggling to understand why farmers need annual tax-funded bailouts, would gain valuable insights if, each time they paid $1.40 for a loaf of bread, they were reminded that the farmer got only 5 cents and the remaining $1.35 went to huge retail and processing corporations."[283]

Other very sensible solutions to the problems average farmers encounter include supporting (and rebuilding) rural communities, encouraging young farmers through inter-generational transfer programs, ending hunger in Canada, and dealing with the terrifying rise of obesity, diabetes and other health problems created by the corporate food system.

134

Perhaps the most crucial item in the whole list proposed by the union – in addition to saving farmland – is the suggestion that farmers turn to organic methods. "Helping redirect farmers from volatile, low-price export markets . . . and helping farmers instead focus on stable, high-price local markets could put billions of dollars in the hands of our family farmers and significantly ease the farm income crisis."[284] If the average farmers turn from the low-price, export markets, as suggested by the union, and move toward sustainable, organic farming, they must also embrace biodiversity. Successful organic farming depends on honouring biodiversity.

Now, because biodiversity depends on a strong connection between farmers and their environment, farmers must become engaged. So, how will that happen?

First, people have to comprehend that we have a problem. A Big Problem. Community-based, ecological-restoration projects will help us find our way out of the maze. Through such projects, community members enter the ecosystem not to see what they can extract or which trees they can cut down, but to look for ways to restore it. As Janine Benyus clarifies in her inspiring book *Biomimicry*, instead of an extractive agriculture based on industry, we can have a self-renewing agriculture based on nature.[285]

By rehabilitating nature, we repair our broken bond with it. How much more peaceful it becomes to live in harmony with nature instead of always trying to dominate it – or one another. Near the beginning of *Biomimicry*, Benyus talks about how much easier it would be if we used nature's models for whatever we wanted to figure out in our own lives. She points out that nature's failures long ago became fossils.[286]

In her conversation about how humans are overburdening Earth, Benyus writes:

Reaching our limits, then, if we choose to admit them to ourselves, may be an opportunity for us to leap to a new phase of coping, in which we adapt to the Earth rather than the other way around. The changes we make now, no matter how incremental they seem, may be the nucleus for this new reality. When we emerge from the fog, my hope is that we'll have turned this juggernaut around, and instead of fleeing the Earth, we'll be homeward bound, letting nature lead us to our landing, as the orchid leads the bee.[287]

Orchids always have at least one petal that reaches outward and serves as a landing pad for pollinators. Their beauty lures the bees to flock to them. In the same way, nature continues to extend a hand, ready to give us the remedy for the global grief we all feel about having destroyed our environment. If we begin to participate in ecological restoration projects, we can reconnect with nature and begin to address our grief.

Farming used to concern a whole community on the land, and it still can. Such a community includes not only farmers and consumers but also soil, trees, frogs, birds, insects and other animals. To stimulate community on a big piece of farmland, a farmer could first explore and preserve the wet areas. Then she could plant alternating thirds of the land with a diverse and complementary selection of crops, amending the soils by adding leaf mulch. During the in-between years, she could allow the soils to take a rest from each primary crop by planting cover crops that enrich the soils as they grow and decompose. In addition, she could plant a selection of trees and native vegetation to promote long-term amendment of the soil and to encourage pollinator habitat. This real-living approach – consisting of wholly caring actions – sustains us all.

Nathalie has often said, "You may think I'm crazy, but when the frogs croak, little things in my ears spin like propellers. Ecosystems are like green hospitals. Some say that when we're healthy, a human's magnetic resonance (or vibrational frequency) is 7.8 hertz, the same as Earth's.[288] When we get sick, we fall below that. You can charge up your frequency using a variety of modalities including Feldenkrais,[289] or, in my case, simply lying in the field and letting either the bees' buzzing or the frogs' croaking raise my vibration. That's the ecosystem benefitting me. We're meant to be cozy with nature, giving and receiving in equal measure – not separated."

Nature constantly extends a hand, offering myriad benefits. As Benyus writes, we need to slow down and study in detail the way nature has worked so well over millions of years, instead of arrogantly assuming that we know better. Imagine slowing down enough to study, for example, burrs and other clingy, sticky seeds, even taking the time to examine them under a microscope. We'd see they have hooks that hold on to soft loops in the fur of animals (or on to our soft pants) so they can be transported to a new neighbourhood. Those who cared to slow down enough and study such sticky things a little over half a century ago were rewarded: they applied the design to their invention of Velcro.[290]

Nathalie tells a story exemplifying how young people naturally want to be engaged in nature. "Because frogs are such a big highlight in my life, I was talking on the radio one day with Jeremy Baker,[291] who was doing a podcast about them on his show, *The Afternoon Zone*. He's a rock-and-roll kind of guy, and he came out to our farm, got into our rowboat and floated out on our pond. He sat still in the boat for a while and recorded the frogs. After he left, I went up to our cozy little treehouse on the hill and fell asleep.

"All of a sudden, lights shone through the windows. I yelled, and David grabbed his hoe as though he were a samurai and said, 'Don't worry. Stay here with our daughter.' He disappeared down the hill, and then there was this awful silence. He didn't come back and he didn't come back, until I imagined he was dead. Finally, he returned. It turned out that two 16-year-olds had been listening to Jeremy's radio podcast on *The Zone* and *The Q.* They were so turned on that they wanted to hear the frogs for real. So, because Jeremy had said where he'd recorded them, they came here.

"Imagine *never hearing* a real, live frog – living and dying and never hearing a frog! This is a true loss of basic culture and part of our loss of connection to Earth. Those two teenagers just symbolized for me the separation of humans from nature and how we really, truly want to be connected."

Ecological restoration is about amending that relationship. Instead of exploiting and cutting, it's about planting and increasing the biodiversity. By committing ourselves to supporting biodiversity, we have the highest chance of fixing our food, health and environmental problems. Once we've connected with nature, we'll want to protect it. By ensuring biodiversity, we ensure food. Food truly hooks us.

Just like Velcro, we want to hook people to connect with each other and with the ecosystem on the farm. Instead of engaging in commercial advertising, Madrona Farm encourages customers to ask, to interview and to observe farming practices in person. The farm is only 12 minutes' drive from downtown Victoria, so it's easy for anyone in the city to reach out and reconnect with the land.

If we could work toward saving farmland close to human communities across each country in the world, we would plant the seeds to amend our broken bonds.

"Cry. Forgive. Learn. Move on. Let your tears water the seeds of your future happiness." —DR. STEVE MARABOLI[292]

SIX : Transcend the Paradox

What Drives the Passionate Activist?

In every community, we find passionate activists, eager to reconnect with nature's biodiversity, eager to amend the farming environment's broken bonds. Knowing the incredible amount of energy it takes to inspire so many people, we asked Nathalie, "What drives you?"

"Hope and witnessing success are a couple. But love especially drives me," she said. "I've had my share of tragedy. My dad, whose passion for nature sparked mine, was absent for what seemed like impossibly long periods, flying military jets while I was young, and he died when I was 17. I've missed him terribly. My mom loved me dearly and has been my greatest teacher. Her pain in life – she grew up during the Second World War and experienced all types of loss – taught me lessons about how society, with its judgments and cruelty, disconnects us all. Surviving a brutal car accident, learning to walk and talk again in my teens, also taught me incredible lessons about reconnecting."

Having experienced Nathalie's vibrancy, we know the average person only sees the health radiating from her now. "I never wanted to be the person who used life's circumstances as an excuse to prevent me from doing anything. I've always needed to challenge myself and maintain hope. I just want to be an ordinary person. So, I search for examples of success as I search for making things right."

Of course, she's always wondered why the world spiralled into such a mess. She's always searching for meaning, trying to find what will make things right. When Nathalie was in Uganda in 2013 for the International Conference of National Trusts,[293] she took a few side trips. During an excursion to see wild animals, she met a woman who said she was glad she didn't have any children so she wouldn't have to worry about the mess she was leaving behind. Nathalie felt very sad for the woman, because she obviously felt powerless. She couldn't begin to dream of living in any kind of way that would help create solutions to the world's problems.

Protecting Madrona Farm in perpetuity showed Nathalie that people really do have the power, that we are a force within the ecosystem. If we're united and respectful, we can make a difference. Because Nathalie and David both had hope, they witnessed 3,500 other hopeful people helping them raise almost $2-million so that The Land Conservancy – not a private family or subsequent investors – could own their farm in perpetuity. The National Trusts, of which The Land Conservancy was a part, are firmly established in 63 countries around the world and preserve both cultural and natural heritage areas (including farms.) They provide another example of a growing potential for lasting solutions.

Embracing the Paradox

We need to keep helping people reconnect with nature and keep sharing our success stories so we have fuel to keep moving us forward. It truly is a paradox: people are both the problem and the solution. Sadly, too many people believe in only the first half of the paradox – that people are the problem.

When Nathalie contemplates the "people problem," she thinks of her son. "When he was young, he was great with the animals – like an animal whisperer – until he was about 7 years old. He used to put dirt inside his boots. When I asked him why, he answered, 'I need to be close to the soil.'

"One day, a huge bully towered over him in the playground, pulling the legs off grasshoppers. My son got so mad that a surprising power welled up inside him. He stood as tall as he could and, in an incredibly assertive voice, he said, 'STOP! STOP RIGHT NOW!' The bully dropped the grasshoppers in amazement, turned and left.

"Because he was so angry, I questioned him. He yelled, 'I hate school! I hate humans because they're destroying animals and the Earth!' It was pretty shocking. On one level, I knew he was totally right, and he was brave and articulate enough to express it. He sounded like a version of those who claim, over and over, 'If we could just get humans out of the equation, the Earth will be fine. Soon enough,' they say, 'we'll do so much damage that we'll produce a virus and all the people will die. Then everything will be fine.'

"All the time, I hear people saying, 'Sometimes, I just want to jump off this poor, messed-up Earth. I feel so despondent about what we humans have done to it.' Now my son is in his late teens. He's pretty much grown up, and like them, he says, 'You've lied to me. Look at this world. It's full of crap. Look at the people and what they're doing.' You know, it's pretty scary to hear your own son carry on like that. I feel so sad about the weight he carries."

Of course, as the insights grow, the heaviness grows in this most passionate and sensitive time of a young adult's life. However, throwing out humans can hardly be any kind of solution. We've all had tremendous opportunities to meet some of the best humans in the world: from compassionate neighbours to kind and genuine strangers, giving a TED talk or speaking to the media. We must meet the people paradox head on and embrace the second half of it.

We have an untapped capacity for greatness. The good, compassionate people we've met can help us turn our taps on so we can create solutions together. We can protect the biodiversity we still have, and we can regenerate the biodiversity that we used to have. If we can't believe this is possible, we might as well jump into our global grief, go swimming with other emotional vampires[294] and drown. Go the other way! Look at the good people who are already leaping into the power of hope. We must listen for the warm buzz of the worker bees. We need to hang out in the shelter of the trees that stand stalwart, providing habitat on the mountainside. Together, as we choose to notice each other's wisdom, we can create a difference.

Overcoming the Bullies

Terrence Webster-Doyle, an inspiring peace educator with a Ph.D. in brain research, learned at the age of 6 how to recognize positive opportunities when the world seemed to be full of bullies. As tall as his Grade 1 teacher at that tender age, he was a frequent target of tyrants who loved to boast about punching and knocking down the "biggest kid on the playground." He never fought back, even when one of their fists knocked out two of his teeth. However, one day when they knocked him down, he landed on a bee that stung him in the back. Screaming in agony, he leapt up to his full height off the dirt, so terrifying the bullies that they ran off the playground. In that moment, Terrence realized that the bullies actually feared him, and he dedicated his life to understanding fear and how we can actually make the choice for peace.

Dr. Webster-Doyle is a scientist who has stepped beyond grief and seized each moment. He encourages everyone to take a breath – or pause to take in the picture – when a bully is trying to engage our reptilian brain and put us into automatic-response mode. Marvin Garbeh Davis's book, *Brave New Child: Liberating the Children of Liberia – and the World*,[295] provides the example of a child living in Liberia in the 1990s. Imagine you are this child and see a person who is all-too-innocently wearing red (the colour of your enemy tribe when you were younger). Instead of allowing your reptilian brain to blindly react, you can take a breath. With that breath, you can realize that *this is now, not then.* You can realize that if you let it, your primitive brain will react according to the old fight-or-flight routine. After you take a breath, you can say "Stop!" to that reaction. At that moment, you can make the choice to engage the more rational part of your brain to take appropriate action: realize that whoever you were initially upset with is not "armed," but "open-armed," and simply say a friendly hello.[296]

Reconnecting to Wisdom

We should take any opportunity to relax and connect with both our own and others' wisdom – notice the bees' buzzing good works, as well as human solutions to problems. When we make the choice to collaborate, we move from experiencing the power of hope to engaging with the even more powerful intention to act. At that point, we can respectfully connect and collaborate with the huge number of answers already available in nature. It is incredibly rewarding when we choose to notice good things, both in ourselves and in our ecosystems.

In *Biomimicry* Janine Benyus gives perfect examples of finding answers in nature. She envisions that "even computing would take its cue from nature with software that evolves solutions" and how "in each case, nature would provide the models: solar cells copied from leaves, steely fibers woven spider-style, shatterproof ceramics drawn from mother-of-pearl, cancer cures compliments of chimpanzees, perennial grains inspired by tall grass, computers that signal like cells, and a closed-loop economy that takes its lessons from redwoods, coral reefs and oak-hickory forests."[297]

Consumers are already taking the first step when they want to know where their food is grown. Groups like Food Secure Canada,[298] the National Farmers Union,[299] the Centre for Science in the Public Interest[300] and the Inuit Tapiriit Kanatami[301] have information websites, are speaking up in Parliament and are insisting on a national plan that addresses food quality, cost and availability. They especially want to connect people in communities with their local farms.

Let's open our ears. Hear the calls. Notice the heroes, both human and nonhuman. We can and we will reconnect.

PART TWO
Take Action

"The frog does not drink up the pond in which he lives." —NATIVE AMERICAN PROVERB[302]

SEVEN : Overcome the Obstacles

Madrona Farm: An Example for International Inspiration

The stories in the first part of this book typify many we've heard from other farming friends and customers at Madrona's farmstand. As we nourished our connections with increased comprehension and felt propelled to take action, we lightened our mutual burden by sharing stories. Increasingly, we realized these stories – and the people who told them – not only followed similar patterns but also revealed a variety of creative solutions that had the potential to help us protect healthy food sources and associated biodiversity forever.

Once people understand how important food security and biodiversity are, they want to protect them. After people have made their first queries and discoveries, they begin to build an ecological awareness, which extends to comprehending why saving sustainable farmland matters. At that point, they have joined a community of people who also feel

strongly about farmland; they have perhaps found mentors –
educators – whose job shifts from one of answering questions
and pointing out resources to helping action-oriented students
climb together over local obstacles.

Each farmer's own local obstacles may at first seem impossible to surmount. Specific problems differ from country to country, from community to community, and from farm to farm. In
every community, creative solutions flourish.

In the Blenkinsop Valley, just nine kilometres (5.6 mi) from
the heart of Victoria, British Columbia, Madrona Farm provides
sustainably farmed vegetables picked fresh each morning, four
days per week, year-round. The Chamberses – with help from
3,500 customers and friends – mounted a successful campaign
to protect Madrona Farm, ensuring it will be a farm forever.
Their model can be used in any of the growing number of countries that have secure land trusts, especially if they are part of
the International National Trusts Organisation (INTO).[303]

The path to securing safety for Madrona Farm was challenging; surmounting the problems took time, money and
creativity. As their determination paid off, Nathalie and David
grew excited to share how they and their friends climbed so
successfully over the obstacles while working to save Madrona
Farm forever.

Obstacle One: Raising Local Awareness about Why Farmland Is Disappearing

Victorians, with their friendly climate and rich soils, perhaps top all Canadians in being loyal to the concepts of the
hundred-mile diet, chafing to buy local, organically grown
food. Farmland around Victoria is also the most expensive
in Canada, actually double the price of the next most costly –
in southern Ontario.[304] In the year 2012, the average price

of Canadian farmland increased by almost 20 per cent.[305] Although single-family home prices in the five years between 2008 and 2013 increased by about 20 per cent in southern Ontario, farmland prices in the same period bloated from 300 to 400 per cent![306]

Lenders, investors and property-savvy newcomers are at the root of the increase in farmland values. Local lenders realize they can make impressive interest on increasingly bigger mortgages. Local investors encourage the boom, renting back to farmers (at best) and/or waiting until the land can be subdivided and sold for housing as suburbs sprawl. Some foreign farmers who want more land realize that they can sell their current property and for the same amount of money move to Canada where they can buy a farm 20 times larger.[307] Other newcomers, having witnessed their own arable land being consumed by mines, dams and real estate, realize the financial benefits of investing in less-expensive, Canadian, arable land. Some farm sustainably. Some farm industrially. Others just hold the land idle until they can turn it over for a profit.[308] Meanwhile, many young Canadian farmers can now afford to pay the mortgages on a few hectares only if they already have high-paying jobs (in addition to their new farming work).

While we are proud of Canada's multicultural mosaic, we could benefit even more by gleaning solutions from the thousands of years of knowledge accumulated by our land's original multicultural peoples. Those who immigrate often assume that Canada's original peoples are ignorant, and the cultures simplistic – without understanding how the land, water, vegetation and creatures differ from those of the old country. They arrive with a mindset understandably full of their own laws, cultures and languages. Local people can easily feel that the newcomers are imposing their own, seemingly inappropriate and disrespectful land use, that they are ignoring natural sustainability and remaining oblivious to the agroecology that Native peoples practised for thousands of years. North and South American history tells these stories over and over again – of newcomers who wreaked havoc on Indigenous cultures and communities. If Native peoples fought back, the newcomers waged war, vigorously and lethally. Now – before it's too late – we must put down our weapons (hardware, chemical, biological, financial and social) and open our minds to a fully shared, mutual education. Only then will we be able to embrace healthy food from healthy sources, shared in healthy physical and social environments.

To overcome the community's general lack of awareness of both the broad issues and the local obstacles, Nathalie, David and their friends made the choice to start a very broad conversation. They have demonstrated that if a small group of people begins to speak and to pick up their pens, together they can be louder than the real-estate investors and the property-tax evaders. Through such conversations and statements, others can begin to understand that industrial farmers have made the choice: to dive into insecticides and to believe that GMO-producing ompanies will somehow not betray them by exerting increasing control over questionable seeds.

The Chamberses, their customers and friends demonstrated another choice: to remain aware and to campaign when necessary. They encouraged everyone to remind governments and corporations that the well-being of both their citizens and their consumers begins with good food. As they talked with customers and neighbours about food security and its benefits, they also suggested that a farm saved in perpetuity would also affect the community in other positive ways.

The Cumberland County Agricultural Conservation

Easement Program in Pennsylvania describes how saving farms has an enticing ripple effect that effectively strengthens entire communities. To quote Vince DiFilippo, the Silver Spring Township supervisor, chairman of an agricultural land preservation board and activist: "When a farm is preserved into perpetuity, home values around these farms usually increase. . . . Saving farmland versus turning it into a residential development actually saves tax money in the long run because it puts less stress on the local school district, does not contribute to traffic problems, helps provide groundwater recharge and requires less road maintenance for municipalities."[309]

Even if governments don't act in the healthiest interests of their citizens, the people on the ground can protect a great deal, especially with the support of secure land trusts. Chapter 11 of this book goes into greater detail about this topic, but the following overview offers a collection of steps anyone, anywhere, can take to save farmland:

- Talk with customers as they come to the farmstand.
- Talk to newspaper, radio and TV reporters.
- Protest at council meetings (at the city, district, provincial and/or state levels).
- Set up a farm website, with links to educational associations, books and other resources.
- Hold fun events at the farm.
- Offer farm tours.
- Hold associated fundraisers.
- Donate some of the farm's food to the needy.
- Ask chefs to inform customers about their local food sources.
- Ask local producers of beer, wine, vinegars, jams or sauces to include sources on their labels and donate a percentage of their proceeds to saving farmland.

Obstacle Two: The Agricultural Land Reserve

In 1973 the province of British Columbia – with the help of passionate farmer Harold Steves (see chapters 1 and 13) – established an institution that soothed the public's concern for farmland protection. "A Commission, appointed by the Provincial government, established a special land use zone to address the loss of 6,000 acres of prime Class 1 and 2 soils yearly in B.C. This zone was called the 'Agricultural Land Reserve.'"[310] The ALR was an entity within BC's Agricultural Land Commission (ALC) for over 40 years. On its home page, the ALR declares: "The Provincial Agricultural Land Commission (ALC) is an independent administrative tribunal dedicated to preserving agricultural land and encouraging farming in British Columbia.."[311] In actuality, the ALC is only quasi-independent, although it was intended to be separate from the Ministry of Agriculture to prevent conflicts of interest.

As we wrote this book, this 40-year-old institution was on the verge of demolition by the provincial government, which had succumbed to pressures from the gas and mining industries. As Nathalie explained to many, "Most people think 'Oh, it's just happening in the North and we don't need that land because it's too far north. But in fact, the Site C dam on the Peace River will get rid of 30,000 acres of the best, prime, Class 1 and 2 soils – food for a million. As the growing zones shift north due to climate change, we will be looking to the northern parts of BC for our food. Thus, it's essential that we use a precautionary principle."

In the fall of 2013, news leaked out that BC's Minister of Agriculture was planning to dismantle the ALR. Nathalie and a group of other highly concerned farmers, scientists, environmentalists, chefs and citizens held meetings and formed the Farmland Protection Coalition (FPC).[312]

Together, the members of the coalition planned a major rally to coincide with the opening of the BC Legislature for the winter session. On February 10, 2014, thousands of irate citizens listened to speakers – including ALR founder Harold Steves, BC's Opposition Critic for Agriculture, representatives of the National Farmers Union and the Island Chef's Collaborative, BC's Green Party interim leader, an independent MLA, farmers from the Site C Dam, Mr. Organic Child Educator[313] and Victoria's poet laureate. Each stressed the need to protect farmland and protested the government's proposal, which at the very least seemed to be a transparent cover-up for promoting natural-gas exploration and development.[314]

Over the following two months, as the premier and her key ministers put forward their proposal to dismantle the northern 89 per cent of the ALR (keeping only the southern 11 per cent,[315] as was outlined in Bill 24[316]), they received a letter from one hundred of BC's leading scientists criticizing "the government for making changes without public consultation or science-based information."[317]

Judith Lavoie of Victoria's *Focus* magazine conducted an interview with Nathalie in May 2014. Lavoie wrote:

> Spring should be an exhilarating season on the 27-acre organic farm. Instead, Chambers is as mad as hell. "I am enraged about Bill 24. Ninety-five percent of British Columbians don't want anything to change with the Agricultural Land Reserve. They want the land protected for future generations," said Chambers, who, like many farmland activists, is angry about the provincial government's introduction of a bill changing the 40-year-old ALR.
>
> Chambers believes the legislation will mean the loss of valuable farmland and she is insulted that the Province did not consult farmers. "Dirt is for real. Civilizations rise and fall on the basis of soil and, especially with climate change, we are cooking up a recipe for destruction," Chambers warned.[318]

As the BC Food Systems Network explained:

> Bill 24 will split the Agricultural Land Reserve into two zones and will devolve oversight of the ALR to six regional panels. Farmland advocates fear changes from Bill 24 will increase the price of farmland for young farmers and will also increase the removal of viable farmland for commercial, industrial and real estate development. This would result in reduced capacity for provincial food security in the face of climate change, as well as increased reliance on imported food, concerns over safe and sustainable agricultural practices in other jurisdictions, and increased food prices due to rising transportation costs.[319]

In order to prevent similar dismantling of legislation that protects farmland in other provinces and countries, it helps to take a look at the ALR's history. The early visionaries who created the ALR had honourable intentions. While Nathalie fought to save the ALR, she was well aware that it was under-regulated, with many loopholes that made it insufficient in protecting farmland and thus food security. These loopholes, weaknesses and consequential losses inspired her to place Madrona Farm with The Land Conservancy. After a year of hearing Nathalie talk about the ALR and The Land Conservancy during the Madrona Farm campaign, most customers grew to understand that the ALR didn't prohibit non-farming use of agricultural land. It simply limited how much farmland could be used for non-agricultural purposes. It permitted farmers to farm land in the ALR, and it was a guiding policy when allocating farmland, but it didn't adequately preserve – or guarantee to reserve – either the agricultural or the ecological values of the land.

In 2006 Charles Campbell wrote an informative and necessary piece for the David Suzuki Foundation, tellingly titled "Forever Farmland: Reshaping the Agricultural Land Reserve for the 21st Century."[320] In it, Campbell points out that although the ALR keeps preserving a basic 5 per cent of BC's land for farming, it constantly juggles the actual pieces of property, releasing some in one location, adding in another. This juggling act has resulted in a severe loss of our most prime farmland. Seventy-two per cent of the ALR land lost over a four-year period ending in 2005 was in the South Coast, where the customer base is largest, the land most fertile and the weather most friendly. Ninety per cent of the added land was in the North, where the human population is low (for now) and neither the weather nor the land as ideal for crops as in the South Coast.[321] However, that northern farmland is excellent for sheep and cattle. With Bill 24's dismantling of the ALR, the new North and Interior regions will be opened to non-agricultural development. Even before Bill 24 passed, the ALR did permit farmland to be re-allocated.[322]

During their campaign to protect Madrona, Nathalie and David would have benefitted from having a few copies of such articles sitting on their farmstand's shelf. Instead, they repeatedly explained that although the Madrona property was part of the ALR, the designation didn't require the land to be kept in food production. Some local farms are run by dilettante farmers who care more about their property tax designation, with its associated tax savings, than they do about actual food production on a reasonable, helpful scale. Nathalie and David knew that the ALR didn't provide enough protection for Madrona, and without more comprehensive protection, Madrona could have been sold, its two designated parcels of land rebuilt with separate homes, converted to a "hobby farm" or gentleman's estate or vineyard, and never again used for growing food for the community.

In other controversial projects in Madrona's Saanich District and in the more western District of Sooke, farmers have unloaded some of their ALR farmland and it has been used for residential development. Likewise, the nearby Municipality of Langford handled at least ten applications in 2013 requesting partial or full removal of ALR land so that developers could request rezoning designations. Meanwhile, more than four thousand individual customers and seven local businesses regularly visit Madrona Farm, annually purchasing many thousands of pounds of produce.

One hundred years ago, after providing produce to local citizens, Saanich Peninsula farmers had enough excess to export to Washington State. Since that time, island farmers have sold much of their prime farmland to developers in Saanich and other parts of suburban Victoria. Today Saanich imports a great deal of its produce from the US, Mexico, South America and China. If anything drastic occurs to prevent these imported goods from reaching supermarket shelves, the small number of local farms will be able to provide the community with food for only three days before running out.[323] If fuel costs rise prohibitively, or strikes or other disasters occur, what will the people eat?

Southern Vancouver Island's weather remains both reasonably warm and adequately moist year round. The soil is of high quality, able to support natural biodiversity. We can reinstate food security, unless we throw away our remaining good farmland. With the safekeeping of our local food sources becoming an increasing concern, we need to realize that the ALR alone hasn't provided enough protection from lenders, investors or the increasing encroachment of spreading urban development.

Farmland supposedly protected by an ALR designation has been slowly dying right in front of our eyes. The ALR looked good at first, but if the community were to rely only on the ALR for protection, it would lose one little bit of excellent farmland after another, until it is too late to save a meaningful amount. The death knell is sounding for more and more farmland.

As Nathalie has suggested, climate change is the elephant in the room. Awareness of the shifting of climatic zones from the south to the north will mean that areas that are currently under the guillotine for oil and gas development will eventually become the bread basket of British Columbia. Does anyone want oil slick as a garnish?

Obstacle Three: Understanding Land Trusts

The Land Conservancy (TLC) was founded in 1997. TLC – formerly BC's National Trust – was modelled after the National Trust of England, Wales and Northern Ireland. In addition to protecting conservation areas, a national trust also takes agricultural and heritage lands under its wing. In the UK, the National Trust owns 200,000 hectares (494, 210.8 ac) of farmland, which are protected and tenanted by hundreds of farmers. Madrona Farm is now also owned by a land trust, The Land Conservancy, and is tenanted and stewarded by the Chamberses. The UK's National Trust is neither a government- nor a corporation-based organization but a membership-based trust. Each member has a voice and a vote. It is a democratic organization.[324]

TLC wanted to use Madrona Farm as a model for sustainable and ecological agriculture to address the largest obstacle to food security on Vancouver Island: the price of farmland. Once TLC owned the property, it would offer long-term, sustainable leases to current and future farmers, following the UK model. The land trust was the commons (see chapter 12) that Nathalie and David used as a mechanism to protect the farmland forever. At that time, no other land trusts in Canada were protecting farmland. When the Chamberses met Ramona Scott, agricultural manager of TLC, Madrona became a conservation partner farm recognized for its care of biodiversity. Ramona had worked on TLC community farm projects before, and she spoke their language. She understood the problem of dwindling farmland, the importance of biodiversity and Nathalie and David's desire to protect their farm forever.

Before Nathalie found TLC, only one of Grannie's sons was on board to protect Madrona Farm. Once she connected with the land trust, she was able to convince the other two. They had the farm assessed. The owners signed a memorandum of agreement with TLC. At that point, Nathalie thought she had done her job and was happy to have helped facilitate the protection of Madrona Farm forever. However, as is explained in the recession discussion that follows, after the first two fundraising deadlines were met, TLC was unable to raise some significant, promised funds from its outside sources. TLC's board of directors had to turn Madrona Farm down. Ramona came to Nathalie and David, saying, "I'm sorry. We just can't do it."

At that point, the whole project hung in limbo. They were left with three choices: they could buy the farm themselves, start their own land trust that could hold a farm, or they could raise the money for TLC to buy the farm. At first, Nathalie thought the best choice would be to start a land trust herself, and she gathered an excellent group of people together. But as her life was already extraordinarily busy, she soon questioned how she would have time to manage a land trust along with everything else, and she decided it would be better to raise the money for TLC.

Because many farmers seeking farmland preservation in perpetuity will be working with trust organizations, to avoid possible disappointment they and their supporters need to know how trusts actually work financially. Basically, land trusts protect the use of land in perpetuity. They acquire land in three major ways: through life estates, donation or community purchase.

For example, landowner Norma Lohbrunner in Langford on southern Vancouver Island did not want her land to be developed. "This is something we always told our children," she said. "They grew up knowing that this home was worth something more than money could buy."[325] She signed a legal, life-estate agreement with TLC that guaranteed she could live on her estate for the rest of her life, after which TLC would preserve her land as a sanctuary for the wild birds her husband had loved.[326] When Norma died in 2011, TLC was given the job of safekeeping her forest and natural, bird-inhabited land.

Marc Theurillat, the owner of Turtle Valley Farm near Chase, in the interior of BC, chose to sign a donation agreement with TLC, agreeing to give his 99 hectares (245 ac) of cattle and sheep pasture, hay field, wetlands, forest, vegetable area, barn, granary and farmhouse to The Land Trust, if it would preserve the property as is.[327]

Madrona Farm, with its vegetables, young orchard, wildlife pond and variety of bird species in a tree corridor, represents the final way TLC acquires land. The Chamberses drew up a community purchase agreement, explaining that their family needed to be paid for the land, but they wanted it preserved in its natural ecosystem state and sustainably farmed forever. After farmers sign this kind of agreement, a community fundraising campaign begins, usually with goals to finish raising the money by a specific time.

Of course, land taken in by a trust may be used and managed for many purposes, including education, but a land trust needs to find the money to buy and oversee the land. TLC has hundreds of volunteer members who help, but it also has a paid staff to handle money, create legal documents, oversee legal conditions that are written up to protect lands, and visit protected properties to make sure they are being treated correctly. If farmers have to pay a lease, staff members must collect it. If a parkland has campgrounds, staff members must keep them in good shape and collect the camping fees. If a garden has a restaurant or store, the land trust must pay the staff and maintenance personnel, and provide funds for food, retail goods and insurance, not to mention keep good books – just like any other business.

Although similar to banks and credit unions in their money management, trusts offer more transparency. Typically, financial institutions of any kind will use deposits as one of their main sources for loans or investments.[328] The dollar you deposit in your bank account may not be the same one you withdraw. For example, when TLC decides to make payments, or establish loans, it clearly accounts for each one of those dollars within TLC's one major pool of money. It would never mortgage a protected property to raise money for another project, just as a bank manager or credit union loan officer would never physically take money from one of our own personal accounts and lend it to another person, or make a loan based on our property and give it to another. However, the transparency of TLC's transfers, as anyone can see in its financial statements, allows those without banking knowledge to access those moves, misinterpret them and decry (if necessary) TLC's decisions. Those with other agendas could also imply wrongdoings, knowing the average reader might easily misunderstand and innocently agree – with outraged horror. Unfortunately, this is exactly what happened to TLC during the period when Nathalie, David and their friends were raising money to save Madrona Farm.[329]

Bill Turner directed TLC for 15 years, during which time he

increased the trust's membership from five to 8,700 members. He had the support of six hundred volunteers, up to 30 paid staff and a board of directors elected by members. During those 15 years, TLC acquired over three hundred large properties totalling 50,585.7 hectares (125,000 ac) all around BC.[330] Securing that much land so effectively required enough knowledge, experience, strength and persistence to take what the general populace might consider risky decisions. Could TLC pay for each? If not, who would? How long would it take? Would that time frame be adequate? How would TLC manage each of the varied properties: parks, campgrounds, farms, heritage buildings with restaurants? The trust also needed to maintain and provide plants for heritage gardens, restore salmon spawning areas in a protected river and keep up a public education program for all properties.

TLC's leader had to guarantee to pay $10,000 to $12,000 worth of mortgage interest payments each month, allocate between $3- and $5-million of annual donations to the correct properties, and manage a monthly operations budget of $250,000 to $300,000 for offices, staff and computer systems in seven different areas of the province.[331] All these concerns demand a leader willing to be criticized for leaping off cliffs when most people would keep their feet firmly planted on the ground. When TLC successfully bought each property, despite a wall of harping naysayers, Bill also had to be prepared to face both appreciation and condemnation. Wouldn't a person have to be foolish to take on such a role? Bill Turner, who received an Order of Canada in 2005 for his work with TLC, responded to this question by saying, "I don't think it's foolish. It's a good way to save the world."[332]

As Bill pointed out, charities depend on the generosity of every individual who supports their work. The largest obstacle comes from those who would undermine a charity's reputation, especially in times of economic recession. During the recession in 2008, some claimed that one of TLC's garden mortgages exceeded its value, when in fact its mortgage was less than one-third of its value. Others wrongfully implied that donors' money was going elsewhere instead of the allocated project. Yet others falsely stated that operating expenses were at least as high as, "if not higher than," donations and other income. False accusations flew around about administrative costs being high when they were actually about 10 per cent of total expenditures (compared to an average of 15 to 20 per cent for most non-profit and charitable organizations).[333]

Bill Turner was fired on March 27, 2009, a gloomy, rainy day. The existing board ran the organization until an election was forced – with votes made by over eight thousand members across BC – in early August of that year. At this point, the first $150,000 had been raised for Madrona Farm. (Nathalie was just beginning to learn how much more the campaign was capable of achieving due to the success of the $100 Climate Action Dividend cheques, which gave many people a perfect framework within which to make a donation.) With Bill's departure from TLC and mounting media reports about the land trust's difficulties, however, Nathalie faced another challenge. Customers arrived at the farmstand declaring, "Oh, Nathalie, we trusted you. Where's our money?"

At first, Nathalie could only say, "Here I was, trying to do a good thing. Now I feel very responsible to my donors and customers for completing the task we set out to do and overcoming any obstacles that stand in the way. I will get to the bottom of this. I'll find out where the money is and what's going on."

As most people found TLC's issues far too complex to understand, they had grown increasingly unsure about donating money to TLC. Nathalie met with Jim Hutchinson, a lawyer experienced in farmland trusts, who kindly donated his services to set up a separate and secure trust account for funds above the initial $150,000 already raised for Madrona's campaign. The trust would allow those funds to be completely refundable if the total, including interest, weren't raised. People could donate through this separate trust, but the money wouldn't actually go to TLC, and tax receipts wouldn't be issued until the Madrona campaign had succeeded in raising the total amount promised. Fortunately, a "white knight" (as the media described her) donated another $100,000, which both boosted the funds already raised and provided positive publicity, inspiring others to donate, thereby helping the campaign succeed the following year.

In the process of meeting with TLC employees and discussing the issues, Nathalie learned a lot of positive information about Bill Turner. She got to know him and realized he was an altruistic, high-integrity visionary who wanted to save the world. But when she went to a TLC board meeting at the White Eagle Polish Hall, Nathalie witnessed one person after another taking the stand and yelling at Bill. The papers from the accountant, which would refute what they all were saying, trembled in his hands. Bill knew he had to face similar attacks in all five TLC regional areas of BC,[334] and watching him standing respectfully, but shaking, she felt a ferocious energy rise up in her that day. She had a lot of respect for him and wanted to protect him. "Bill," she said, "I'm coming on that road trip with you."

Recalling their trip around the province together, Nathalie said, "I learned how humble and brilliant he is. Above all, I witnessed his altruistic ethic for nature and for farmland. It made me realize that a lot of people see greatness in retrospect. I couldn't stand around idle while Bill got creamed when I knew then – and now – that he's remarkable. So, I volunteered on the Save TLC campaign."

She phoned hundreds of TLC members to encourage at least the required 70 per cent of the eight thousand members to vote Bill back in during the election of the new board. During this campaign, Nathalie heard people say, "Oh, Bill's great." But she also spoke to some who said, "Oh, I don't know about him," and she began to lose faith in her ability to rally support.

She recalls the moment clearly. "Bill was sitting in a café, wearing jogging pants and looking super stressed-out," she said. "I admitted to him that I didn't think it would work, that we'd better look for a Plan B."

Bill looked at Nathalie and said, "If we do not quit, we will not fail."

"At that moment," said Nathalie, "I felt like a frog on a lily pad and made a conscious leap of faith. I manufactured the faith to make the leap. Although I hadn't believed we'd succeed, I also realized I didn't want to live my life jaded or in doubt. So, I decided that even if I didn't believe we'd succeed, failure would be too huge a calamity. I had to believe. So, I just carried on."

A chief financial officer was brought in to replace Bill.[335] Some of TLC's original staff began addressing the accusations through the formal, protocol process of organizing requisitions – petitions – in order to gather signatures from at least 10 per cent of TLC's members to demand an extraordinary general meeting (EGM). At the EGM, to be held by the end of June 2009, the board would have to report on changes and new developments, then discuss and vote on a special resolution to formally replace the directors. As a result of this process, they were also hoping to clarify the public's perception – a tall order, given the complications of the board's process, the technical details of any trust's banking and the mortgage processes between TLC, the District of Saanich and David's uncles. Assuming the petitions granted the EGM permission to go ahead (and assuming the board succeeded in working through the clarification process), it would all culminate in another vote from all TLC members for a new board of directors.

Meanwhile, accusations and reassurances bounced around. The media published twisted views. Two sides formed, one called TLC Forever, the other called Save TLC. Madrona's farm kiosk quickly became more of a political stand than a vegetable stand.

On a day when the media had arrived to publicize Madrona having reached a fundraising deadline, the parking lot filled with cars, people shopped for vegetables and the two political sides took the opportunity to advertise their viewpoints. When the TLC Forever group began handing out leaflets against Bill Turner, Nathalie asked them to leave. The media leapt on the drama. "Please don't show this on the news," she begged. "I've asked you to come to help us celebrate reaching our second deadline, not to broadcast a dispute that will turn other donations off."

Soon, as David said, "all the veggies came with drama – actually a lot more drama than vegetables. Business definitely went down for a while. We set up tables and used the barn for TLC board meetings. The barn was almost like a war room, except that swallows were all nesting inside it over the course of the summer, and people couldn't help commenting on them. The time came when babies were born, and mother swallows flew in and out bringing food. We finally had to put cardboard on the tables because the swallow babies were always pooping. By the time the last requisition meeting happened, with a barn hanging full of our freshly grown garlic, there was no room left for meetings."

At last, the requisition process and board meetings ended, and a little over four months after his ousting, Bill was reinstated in a landslide victory on August 8, 2009. Nobody from the other side was voted in. "It was a miracle," Nathalie said. "I became a believer."

"At that point," she said, "I knew that Madrona Farm would be protected, and I realized how democracy – from the power of each TLC member casting a vote – created a real power of the people. In direct contrast to the media hype by those who would like to twist and contort information, the membership strength of those who took the time to truly inform themselves really inspired me."

"But you know," said David, "in the long term, this kind of victory probably wasn't the best, because it basically polarized the community."

David's comments refer to a distressing amount of negative targeting of TLC. When fabricated reports started circulating about its revenues falling flat in 2010, TLC had to take the time to point out that despite an uncertain economy and adding 1618.7 hectares (4,000 ac) of land to its collection, 2010 revenues had grown by 7 per cent over 2009.[336] Some people falsely accused TLC of using a line of credit to pay a matching donation, so TLC had to spend time showing that this was not the case. Some, who were disappointed that they could not continue using formerly private lands in a way that would disturb the new public trust park visitors, resorted to myriad threats at all times of the day and night.

Although all people, wealthy or poor, are equally capable of prejudice, those with more money and media connections have the advantage of being able to tell their stories and interpretations of events more broadly. When this indeed happened, TLC was the target of a whisper campaign – criticized for saving properties that disappointed businessmen had hoped to develop. Competing trusts could likewise complain about TLC ever so politely, followed by a plea to donate to their own cause which, with careful wording, sounded more secure. Others could simply use the phrase "financial situation" in a disapproving tone when referring to Madrona Farm, implying that it was in jeopardy, when in fact the money had been raised and the appropriate papers were completed to protect the farm.[337]

Due to such pressures, in the fall of 2012, after many of TLC's knowledgeable staff had been let go by the board, or on orders from the board – and others left when they saw what was happening – implications that TLC was facing bankruptcy hung heavily over everyone's heads. Earlier, in May of 2012, the board had given up on the protection of historic and farm properties, deciding instead to embark on a nature-only approach. Within six months, the board faced the fact that TLC could no longer pay down the principal but only the interest on loans – and they gave up on fundraising. TLC's staff had to take more time to explain to the worried public that bankruptcy insurance is a smart item to have in any protection portfolio, and even if anything should happen to TLC, the properties that were under the trust's umbrella would remain protected under provincial law through the Charitable Purposes Preservation (CPP) Act. The reorganized board members wanted to sell part of Abkhazi Garden, Binning House, Wildwood and Keating Farm. In the end, by going to the court under creditor protection, they were able to get around the CPP so they could (and did) sell Keating Farm.[338]

Somehow, Bill Turner survived the highs and lows of creating and growing TLC. He said, "Well, I think we all have to be able to laugh at ourselves and have fun. We've always tried to have fun here. It's often hard to party or celebrate because we're on to the next project, but it's really important to have those parties or special lunches that we try to have."

A "positive outlook" kept Bill going for the 15 years of his contribution to the birth and growth of TLC until his retirement on June 21, 2012: "A belief that if you set your mind to something, you can do it. I think that's something that's been part of this whole team and has been there from the beginning."[339]

People asked what he'd miss the most as he chose to move on from TLC to centre his work within the International National Trusts Organisation (INTO)[340] a year later, after receiving the Queen Elizabeth II Diamond Jubilee Medal for his hard work.[341] He said, "That's a really hard question. What I'll miss the most is the team, and the *can do* attitude of the team. And the inspiration that comes from that team."[342] His new organization is built in a way that makes it abundantly clear that agriculture and historic properties are a key part of the mandate, and that the role of its trustees is that of protecting these special places forever.

When he was asked to share advice, Bill referred to inspiration from Michelangelo: "There's always been more risk in aiming low and knowing you will succeed than aiming high for the sky and hoping and knowing that you can succeed as long as you don't give up. I hope everyone here will remain confident and positive in knowing what they can do."[343]

Bill's words of wisdom are appropriate for all of us, as the challenges associated with saving farmland will continue. However, we hold the strength inside ourselves to rise to every trial.

Obstacle Four: Recession

Deadlines, although great for achieving goals, seem all too often made to be broken. David and Nathalie set three incremental deadlines for raising money to save Madrona. They raised the first $250,000 by December 3, 2008. The next $500,000 needed to be in place by July 31, 2009. Reaching a total of $750,000 on time, they felt reasonably confident about reaching their final goal of $1.7-million by January 30, 2010. Some people say that things happen in threes. In this case, these three financial goals were stymied by three obstacles.

TLC staff said, "I'm sorry. We can't do it. We can't raise the money we thought we were going to be able to add to the fund. The recession is too bad." Teck Resources Limited, a Canadian mining and mineral development company operating in the Kootenays, was feeling the consequences of the freeze and had withdrawn their promised $5-million donation to TLC which, over a five-year period, would have provided an endowment fund, an internship program and working capital. TLC staff had planned to donate $250,000 of this to Madrona. Second, TLC requested an additional 10 per cent, or approximately $250,000 to care for the property in perpetuity. Finally, David's uncles had the farm reassessed, adding another $500,000 to the fundraising goal. Not only would Nathalie and David have to extend the final deadline but they'd have to raise an additional $1-million.

When a year had passed and TLC still could not materialize anticipated donations, Nathalie started a new land trust board with paleobotanist Richard Hebda,[344] and soil scientist Bob Maxwell.[345] They completed the paperwork for their new Saanich Peninsula Farmlands (SPF) Trust and talked with the Land Trust Alliance, a union of over two dozen community land trusts within BC, in addition to the two province-wide land trusts that include The Land Conservancy. Throughout the early parts of their TLC campaign, Nathalie was not only coping with being a new mother but was also studying First Nations history and ethnobotany at the University of Victoria with Nancy Turner.[346] Her whole life felt as though it were turning upside down as she became immersed in protecting farmland. She wished she had the super powers necessary to somehow follow through with the proposed SPF Trust, but she was exhausted.

On the bright side, the major recession really helped educate potential donors about the precarious level of food security. People really started to understand the contrast between the robust food security of one hundred years ago and today's paltry three-day supply. The recession made it easier to imagine a time when the ferries might not be able to run and islanders would need access to their own, functioning farmland.

Perversely, the recession may have compelled the need to donate, and donate seriously. Many decided to donate their $100 Climate Action Dividend cheques. Although the original January 30, 2010, deadline had to be extended a couple of times, a community of thousands continued to rise. Donors ranged from elementary-school students to seniors' centre participants, from hockey-pool winners to Earth Month participants, from hotel staff to restaurant customers, including Canadians, Americans, Britons and Singaporeans. Madrona's diverse campaign succeeded in climbing over every obstacle to raising the final total by mid-spring.

Obstacle Five: Navigating Family Relationships

Families often reflect a balance of personalities, views and preferences. While one person may let frustration fly as freely as sparks from a match, another may remain calm as a crescent moon. No matter how antagonistic one person's actions may seem at any given time, it's usually possible to dig up his or her corresponding positive points. Positions on each end of this teeter-totter may slide in or out with time or with counselling, but the overall balance within the family usually remains. On a greater scale, the same often applies within – and between – nations.

In the Chambers family, one, high-pressured uncle wanted all his inheritance from the Madrona property as soon as possible. Another uncle, who was easy-going, and David's sister both wanted to donate a major part of their legacies to TLC. David's father and, of course, David wanted to give all of their proceeds to preserve the family farm forever. Various cousins sprinkled their desires along the continuum in between.

Tensions raced between them all, especially when, after a year, the memorandum of understanding between TLC and the brothers seemed to be falling apart. Nathalie suggested that the Friends of Madrona Farm could raise the money for TLC. When TLC insisted on another 10 per cent for staff to manage the property in perpetuity, the more agitated uncle insisted on a property reassessment, just as Victoria's housing market was reaching its peak. The value of the back piece of property had increased from $1.4-million to $1.5-million. Crowning it all, the 1.4-hectare (3.5-ac) piece of empty land fronting 20 metres (60 ft) of road provided a straightforward housing construction opportunity, so reassessment of that part of the property resulted in its value leaping from $600,000 to $1-million.

Obstacle Six: Confusion between Farm Projects

Just as Madrona launched its campaign, competition reared 19 kilometres (12 mi) farther north on the Saanich Peninsula. Woodwynn Farms was also throwing its fundraising drive into gear, aiming to help the homeless redirect their lives.[347] Upon hearing that "There's a farm being protected," a potential donor would open her wallet. Nearby, another would say, "Oh, which one are you giving to?" The public became confused about which farm was which.

That year, as usual, Madrona Farm set up a display tent at an annual celebration called the Feast of Fields.[348] An organization called FarmFolk CityFolk (FFCF) imported the idea of the Feast of Fields from Ontario to BC in 1995.[349] It encourages city folks, farm folks and chefs to have fun increasing their connections with one another and to their food's ecosystems. While feasting, participants also fundraise for healthy food boxes, sustainable farming, chefs' collaboratives, community gardens and many other associated, food-related causes.

As Nathalie drove up to unload supplies for her display tent, she suddenly gripped the steering wheel and debated whether or not to climb out of the van. The Feast of Fields had allocated Madrona's display to be right next to that of the Farmland Trusts (FLT), which was fundraising for Woodwynn. Barbara Souther, the founder and chair of FLT and Woodwynn's actual next-door neighbour, looked curiously at Nathalie's vehicle.

"Where did you get that van?" she asked.

"Oh, from a friend who was selling it for the owner," Nathalie said, as she stepped down.

"How funny," Barbara said. "He's my friend, and that was my van!"

Before long, Nathalie and Barbara were shrugging their shoulders and saying, "We should work on this together." Their friendship blossomed.

While Nathalie promoted Madrona, Barbara worked for Woodwynn. However, in a twist of events, in the early spring of 2009, another charity – the Creating Homefulness Society – won the right to purchase Woodwynn.

Just as TLC staff and Madrona were riding the roller coaster of Bill's spring firing, and a new CEO had been hired in his stead, Barbara Souther was scrambling to locate funds that she had raised for a covenant on Woodwynn that TLC was holding in the amount of $200,000. Although Barbara was devastated at the failure of the Woodwynn project, she rose to the highest level of philanthropy as a true defender of conservation and insisted that donors who had donated to the Woodwynn project but had not yet been issued receipts, redirect their funds to the Madrona campaign. Some of those donors had been issued tax receipts while others had not. Barbara worked with individual donors and helped channel the money toward Madrona Farm.

On a day when Madrona had just met one of its fundraising deadlines, Nathalie was talking to Ed Johnson, a Woodwynn neighbour and future chair of the Farmlands Trust, who had changed the direction of his donation from Woodwynn to Madrona. He said, "So, what's your great plan, Nat? How are you going to raise the rest of the money?"

"Of course, I'm searching for Jimmy Pattison [a Canadian philanthropist, who in 2013, with $9.5 billion, was considered the richest man in the country[350]]. I have this idea that I want to create the Jimmy Pattison Corridor of protected farmland running down the whole Saanich Peninsula," she said.

Ed laughed his head off. "If Jimmy Pattison donates $300,000, I will match him."

They were just joking of course, but Nathalie soon had an epiphany. She woke up in the middle of the night, thinking, "That's it! I'm going to ask Ed to advertise a challenge, so others will match it – and maybe we'll raise double the amount being transferred by the FLT." So, instead of the Jimmy Pattison Challenge, they created the Ed Johnson Challenge. To help raise money for the next deadline, Nathalie arranged to create a YouTube video, featuring Ed Johnson in casual – not FLT chairman clothes – stating that he would donate $200,000 to Madrona if the public would match that amount. One of Nathalie's many talents is her ability to find just the right person to do a job. This time, it included a videographer capturing Ed on Madrona Farm, promising the donation in the form of the perfect – and, it turns out, successful – fundraising challenge.[351]

One time toward the end of the campaign, when a truly exhausted Nathalie felt like giving up, David took the family on a quick trip to Galiano Island. As they were walking among the arbutus trees, admiring the views, enjoying the birds flitting about in Bluffs Park,[352] which David's grandparents had donated to the island, they came upon the plaque erected by the island in thanks. "I suddenly felt Grannie backing me. Action dispelled despair," said Nathalie.

"Change is the essence of life. Be willing to surrender what you are for what you could become."

—REINHOLD NIEBUHR[353]

EIGHT : Choose a Model

Components of the Madrona Model

The Madrona Farm model addresses the largest obstacle to food security on Vancouver Island: the price of farmland. With 70 per cent of the island's farmers retiring in the next couple of years, with mentorship possibilities dwindling and with farmland being consumed by both urban and industrial development, this model removes the farm from private ownership and places it into trust. Here's how it works:

1. The land trust owns the farm and offers long-term, sustainable leases to farmers. The land trust then makes sure farming purposes and practices are in line with those agreed upon by the original owners and the land trust. The conservation organization maintains control, and the agricultural and ecological value of the land is maintained.

2. The long-term leases are offered to farmers at market value ($150 to $350 per 0.4 hectares [1 ac] per

year, depending on soil quality for the going rate). Rent is paid to the land trust to care for administrative costs. This is a win-win solution, addressing the unaffordability of farmland as well as the fact that some farmers who are able to secure only short-term leases without tenure are stuck trying to decide whether to invest in perennial crops without knowing if they'll be farming the same land the next year.

3. In the Madrona Farm model, the current farmer must note the sweat labour contributed and improvements made to the protected farm, so the farmer that is next in line for the lease can then buy the pieces of equipment and seeds. The ecological and agricultural value of the land is protected in perpetuity. Leases are drawn up between the land trust and the farm tenants, and in order to stay on the land, they must be farming as laid out in the lease: farming under the principles that the trust has specified.

In the future, some may call Madrona Farm a "community farm," while in fact instead of having many shareholders, it is a family-run business with technically only one operating farmer, and customers instead of shareholders. It cannot escape its identity as a community-supported entity, however. And nor would it want to! Friends, customers and philanthropists are the foundation of both the farm's and the campaign's success. A diversity of sources for funds is the key to successful fundraising: "Don't put all your eggs in one basket!"[354]

In the spirit of diversifying, Nathalie is working on solidifying a Chef Survival Challenge template that can be tweaked to work across the southern end of Vancouver Island. This community event is a large part of the Madrona model. It began as a fundraiser for the Madrona campaign in 2008. Local chefs hang, hurdle, balance, zipline, climb, crawl, and race their boats to "condiment island" and then race around the farm to forage for the vegetables they need to create masterpieces on camp stoves, which they then share with those who bought tickets for the event. "Eventually, I'd like to see Chef Survival Challenges taking place all across Canada," says Nathalie.

"The Chef Survival Challenge reconnects people to the land," she says. "Farmland is our heritage, and it makes up who we are." Many of us are connected to a farm, through some memory or other, maybe spending summers on Grandma and Grandpa's farm. Perhaps we happily reminisce over the smell of hay, or falling asleep to a cricket chorus. Many of us hold such memories, but the last two generations are different, having grown up during the influx of industrialized farming. We have lost a farming culture in a short space of time.

As part of an age-old tradition, some landowners were cutting trees down on Saanich farms (for a coming tax review and farm assessments) in the early 2000s, but Madrona kept its trees because they define one of the major foundations of biodiversity even if, sadly, they are not yet typically seen in farming practices. After their neighbours cut down their trees, David and Nathalie felt squeezed and stressed. They took solace in realizing that at least the two of them were travelling on the same track, completely committed to biodiversity. Thankfully, the provincial Farm Assessment Review Panel in 2009 rewrote "rules that had initially created a tax incentive for farmers to cut down trees."[355]

The Chamberses' practice of promoting and supporting biodiversity includes being aware of other projects going on in their vicinity. Nathalie has done more than focus solely on her own land, and she has spoken out against the astronomical disadvantage to young, up-and-coming farmers. Over her years of farming and fundraising, she has donated money to projects whose actions model biodiversity in brilliant ways. Some of these include the *Victorian Naturalist* journal,[356] Food Roots,[357] Share Organics,[358] Life Cycles,[359] the Farmlands Trust,[360] Pender Island Community Farm Project,[361] and Horse Lake Farm.[362] The list grows as Nathalie's awareness stretches.

Enhancing the Madrona Model

The ever-changing nature of politics and the ALR, land trusts, other organizations and people make it difficult to pin down one, solid way that will work to save all farmland. The Madrona model presented in this chapter worked for Madrona in its specific location and during a certain period in time. Some aspects of this model may not work in other situations, but many could be used (some with alterations) in future campaigns of a similar nature.

In the interest of conserving farmland, Nathalie urges that we take the Madrona model a few steps further:

1. We need to be able to have covenants on farmland. At the time of writing, it was difficult to place a covenant in the ALR. Even when land is within the ALR, it is at risk of being developed, especially as many people buy farmland on speculation, anticipating the possibility of removing it from the ALR. One precedent makes room for many more. Developers and agrologists get together and "prove" that the soils won't grow anything. However, soil is alive and can be improved by farming practices. When humans get sick, we don't ask them to lie down on the ground and roll concrete over them. We treat them with care, improve their health, provide them with better nutrition, suggest eating food with specific vitamins and minerals. The ALR regulates certain uses: developing the

land for subdivisions isn't allowed, but other acts that basically ruin the agricultural and ecological capacity of the land are allowed. David and Nathalie have been protesting municipal ALR deletions for a long time. Covenants associated with farmlands could be similar to those in the American model. In the US, massive strips of farmland are protected through the Purchase of Development Rights (PDR) (see next point). This would take the high price tag off farmland and emphasize its actual worth as essential food-producing property, drastically lowering its cost.

2. If the Agricultural Land Commission would allow covenants in the ALR, land trusts could buy the covenants and the development rights. This is a technique used by the American Farmlands Trust. It is called the Purchase of Development Rights and has protected millions of hectares of farmland in the US. First, the farm needs to have a professional assessment to determine its market value. It is a perfect way for a farming family to protect a legacy and still be paid for its farmland. Please check out the American Farmlands Trust website for more information.[363]

3. If we look at the fact that 70 per cent of the farmers on Vancouver Island are ready to retire, and 90 per cent of them don't have a succession plan for their farms, it's easy to see how the next farming generation is at disadvantage, especially considering that these farmers often have more than one child, and each one wants a piece of the pie. Providing money to children often brings farm owners to the ALC for applications to remove land from the ALR, sometimes more than once. Succession planning is essential to generational farmland transfer.

The Madrona Model and the National Trust

After the Madrona campaign wrapped up, Nathalie started working as the agricultural program coordinator at The Land Conservancy, helping 90 farmers throughout BC. Also working with the Conservation Partners Program (CPP) started by Nicola Walkden, Nathalie reinforced her original connection to TLC. CPP recognizes farmers who are making a contribution to biodiversity and stewardship on farmland. Working with the CPP has been a fulfilling way for her to spend a little time observing events from the other side of the table.

"Working with farmers is really important for protecting wildlife habitats that are endangered," says Nathalie, "because, in BC, most of the conservation value must include wildlife corridors with their species at risk and their endangered plants, because it all affects the success of farmland." Nathalie has met with farmers and excitedly helped them assess their farms and plan for even greater sustainability. Those farmers who were making a difference – who, for example, weren't cutting down their stands of endangered Garry oaks – were given a sticker label and profiled on TLC as a conservation partner.

Madrona Farm naturally became a conservation partner as it developed into a sort of gateway channel for succession planning.[364] Nathalie also discovered the International National Trust Organisation (INTO). This trust holds land, and the money used to purchase it, while leasing it out to people who will steward it according to the trust's quality requirements.

The National Trust for Land and Culture, British Columbia, started up in the spring of 2013, shortly after TLC took agriculture off its mandate. "It is going to be the prototype for a Canadian national trust," says Nathalie, "on Prince Charles's suggestion." While waiting to perform a review of the best way of going forward across Canada, the organization is still acting as a farmland bank.

INTO connects national trusts around the world.[365] At this writing there are 65 national trusts, with a membership of 6.5 million people and growing. INTO has the ability to represent countries whose governments are not working in the best interests of conservation heritage and agriculture by working with non-profit, community organizations instead of governments. National trusts are membership-based, so each member has a vote. Therefore, INTO allows its members to take part in directing a conservation movement.

When the INTO conference was held in Victoria in mid-October 2011, Nathalie worked as a TLC employee giving a farm tour to delegates from all over the world.[366] Although not directly involved with the INTO event, she became so excited about the goings-on that she kept pestering Bill Turner, then head of TLC and organizer of the conference, to let her attend. He kept saying, "Well, I don't know . . ."

As Nathalie tried to find her way into the conference, Bill finally found her a job: delivering papers and chairs to the convention centre. She arrived at the event, started unloading and gasped in amazement at the crowd of people who had come together from all over the world: Africa, China, India, Korea, Taiwan and Australia. As she was desperately trying to figure out how to dive into conversation with them, a group approached her, asking where they could get a bite to eat. Nathalie certainly knew the answer to that question, as Madrona wholesales produce to some of the best restaurants in town. She introduced herself and took them up to Vista 18 – a restaurant specializing in international dishes, located in the tallest building in Victoria – where Madrona's produce is served in style by chefs who look out their top-floor windows to Madrona Farm in the distance and say, "That's where your vegetables have just come from."

As they stepped into the elevator, Nathalie recognized Victoria's mayor Dean Fortin. She introduced him to everyone. The chefs took time to sit with them in the restaurant. That entire weekend, it seemed that she always arrived at the right place at the right time.

Nathalie fell in love with the people she met at the conference; common threads of cultural heritage and agriculture wove into their braided conversations. As events unfolded perfectly, Nathalie remembered the strength that perseverance provides, even when we can't, at first, see its results.

A national trust such as Canada's differs from any government-sanctioned group, not only because it is membership driven but also because it protects heritage, culture, natural areas and agriculture.

When Nathalie visited the National Trust of England, also in 2011, she dreamed of spreading the word about Madrona's model and agricultural principles that deal with ecological protection, soil, culture, conservation and biodiversity to all the other national trusts around the world. However, when it came time to submit her paper for the September 2013 Ugandan INTO conference, she didn't see agriculture on the bill. She already knew that not all national trusts were promoting agroecology, but she hadn't expected such devastating news.

Being an A+ optimist, Nathalie wrote an abstract that would fit in at a conference that did feature agriculture and sent it off. She enumerated how to survive in the wake of the Green Revolution and degraded ecosystems. She pointed to agroecology as the bridge to bring us back to traditional and sustainable farming systems.

A Conscious Culture of Conservation

Like many of us, Nathalie loves Canada's forests, mountains and trees. She admits, though, that she doesn't treasure the more common definition of "culture," which stresses socially shared belief patterns stemming from families or nations. Instead, she is a worshipper of the culture of the ecosystem.

Canada is a relatively new political entity, compared to places like England and France, and its heritage buildings are also new. When we think about the legacy of our country, we often think of mountains, the oceans, climate, northern heritage and the various Indigenous intelligences: truly Canadian landmarks, philosophies and ways of life.

"In order to preserve this," says Nathalie, "we need to develop a conscious culture of conservation. In terms of farmlands that we want to pass on to future generations, we need to use the precautionary principle and draw upon the wisdom of experience from other countries around the world that have already developed and redeveloped and saved what they had left."

We need to be thinking of preservation first and foremost in terms of agriculture. Part of Nathalie's obsession with farmland relates to her Madrona connection, but she also claims a sort of cellular memory of one generation.

"I've always had an attraction to farms, and it wasn't very long ago that my aunt was telling me stories of the family farm on our quarter section in Bonnyville, Alberta, where my dad was born," says Nathalie. "Those stories didn't contain all the dirt and sweat but were all about the gems. When I arrived at Madrona, and I connected with the land, some of those memories started taking over, and I bonded with Madrona really quickly. This connection and my conservation ethic made me realize that this property needed to be protected as a legacy piece for future generations, so I tried to figure out a way to protect it." At that time, TLC's Ramona Scott was looking at protecting land in perpetuity for future generations, as well. She understood the ways in which agroecological and biodiverse farming affect sustainability, so she and Nathalie connected easily.

The Transferable Lease

When Nathalie first sat down with her lawyer, Jim Hutchison, he asked, "Why are you doing this? Why do you want to protect a farm and forfeit your inheritance?" Nathalie launched into her story of how she used to sit by the lily field or the great oak tree and look up at the sky and feel a connection to nature beyond anything she'd felt recently. She calls it a "premature conservation ethic."

As Nathalie expressed her childhood affinity with the Rogers Farm, which has since been consumed by subdivisions, Jim started to shake his head. "You know what," he said. "I'm really sorry, but I think my family is responsible for that." The grandson of Bruce Hutchison, who purchased the Rogers Family farm almost a century earlier,[367] Jim really understood the campaign and leapt at the chance to support the model.

Jim created a transferable lease to go with the Madrona model. A market-value lease, it protects the farmer and will be passed along with the farm. Jim also opened up a trust account for the campaign, so everybody's money went into that account and accumulated interest for the duration of the campaign. If the goal wasn't reached, money would have been returned, with interest, to all donors.

The lease is also replicable. For the first time in BC, a farm was protected for food security for future generations as a result of a community campaign, where the community raised the money. Jim modified the template for TLC and also made sure it protected the Chamberses as tenants. They settled on a 29-year lease with a renewal clause; if they farmed longer than 29 years, they would have to pay part of the principle. (They would have to pay 2 per cent of the property value, so it would be $50,000 to secure the lease for 50 years.)

Just as the lease is a template, and repeatable, so is the model, the fundraising approach and, thus, the campaign. The Madrona model differs from other ways of saving farmland in that the community supplies the funds. Similar projects may source funds from a variety of places, including partnerships between municipal, regional, private and land trust bodies. In the Madrona model, none of the money comes from the government. The land trust simply holds the land. The farmer leases the land at market value, so he or she is responsible for everything, including property taxes. Its bylaws state that if TLC should become bankrupt, the farm would be transferred to another, similar organization, where it will remain forever protected under the Charitable Purposes Act.

When the next farmers assume the lease, they will buy their business from the Chamberses. Nathalie and Dave have no equity either in their farm or in their house. But their equipment – the $70,000 underground irrigation system, the five tractors, the hydrants – is theirs and saleable. As the model includes a succession plan, the next farmer doesn't have all the start-up costs but instead simply rents the land and buys the existing equipment, making the overall price reasonable. The Madrona model addresses the biggest obstacle to food security on Vancouver Island – the price of farmland – by removing it from the equation.

If a farmer buys a piece of land for $800,000 at a borrowing rate of $1,000 per month, for every borrowed $100,000, she's buying a pretty average slice. Let's say she's really lucky and can afford $100,000 for a down payment. The balance in the form of a $700,000 mortgage means monthly payments of $7,000, plus $40,000 per year in start-up costs, unless she owns a tractor and irrigation system, already has seeds, has free water and volunteer farmhands instead of employees. For farmers interested in delving into specific costing, BC's Ministry of Agriculture and Lands provides a clear budget model called "Planning for Profit." Their 2008 example gives detailed specifics of expenses and income for a "Five Acre Mixed Vegetable Operation: Full Production."[368]

In the Madrona model, the Chamberses pay a percentage of market value per month and make their own business choices, such as how many farmers work with them, how to protect themselves from blights and when to open the stand for sales. On a rare summer day, during the height of production in July, the Chamberses might bring in $1,000 selling vegetables. However, most farmers' daily incomes are not consistent.

It would be arrogant and foolish to think that one model could be the solution to all the complications besetting the world of sustainable agriculture and the striving for food security in Canada. In the spirit of biodiversity, we recognize that the Madrona model is one among many, but it has withstood many challenges. Its foundation is strong because of its reliance on community support.

Other Farmland Models in BC

If we look around, we see activism that reveals itself in social organizations brimming with equal parts discontent and hope for the future. Examples of other models include different types of co-operatives, communes and Community Sustained Agriculture (CSA) farms.

CSA farms can include co-ops. These farms are beautiful,

organic and supported by local citizens who pay in advance for their weekly boxes of seasonal produce. As Anne Warren of Notch Hill Organics, BC, says, CSA "is a program of mutual commitment between a farm and a community of supporters. It links people directly with their sources of food, with an emphasis on fresh, local and sustainable. It gives farmers a source of income at a time of high expense (early spring). Supporters purchase a share in the season's harvest and receive this as a weekly package of produce."[369] Each CSA farm is unique. A lot of impressive and diverse action is tucked away in rural pockets all over British Columbia.

Frank Barrie, editor of the website Know Where Your Food Comes From, says Community Sustained Agriculture

> is a growing movement in the United States and Canada, which has provided economic support and predictability to farmers, practicing traditional family farming, that they will be able to continue to farm, and that communities will be able to enjoy fresh, local farm products. Partnering with local farms by purchasing a "share" in the season's harvest, consumers can know where their food is coming from and have the satisfaction of supporting local agriculture. Participants, who purchase their shares early in the season, provide the farmer with a stable income, and in return they receive a weekly supply of fresh local farm products.[370]

FarmFolk CityFolk (FFCF) is a not-for-profit society located in the Vancouver area.[371] It collaborated with Simon Fraser University and The Land Conservancy in 2008–2009 to create a study on the community farm model as it exists in a variety of situations in British Columbia. One of FFCF's board directors, Cory Pelan, the owner of the Whole Beast[372] (an artisan delicatessen) was an immense support during the Madrona campaign. FFCF also keeps an online directory of approximately 30 of BC's CSA farms.

We invite BC readers to explore examples of many CSA, community, and co-op farms in British Columbia, including these:

- Alderlea Farm CSA, Duncan[373]
- City Farms Co-op, Vancouver[374]
- Foxglove Farm, Salt Spring Island[375]
- Fresh Roots Urban Farm CSA, Vancouver[376]
- Glenn Valley Organic Farm Cooperative, Abbotsford[377]
- Glorious Organics Cooperative and Fraser Common Farm Co-op, Aldergrove[378]
- Green City Acres CSA, Kelowna[379]
- Haliburton Farm, Victoria[380]
- Horse Lake Farm Co-op[381] and CEEDS (Community Enhancement and Economic Development Society),[382] Horse Lake
- Lawns to Legumes CSA, Kelowna[383]
- Makaria Farm CSA, Duncan[384]
- Nathan Creek Organic Farm, Langley[385]
- Notch Hill Organics (formerly Sudoa Farm), Sorrento, BC [386]
- Ohm Organic Farm at Yarrow Eco-Village, Yarrow[387]
- OUR Ecovillage, Shawnigan Lake[388]
- Providence Farm, Duncan[389]
- Soil Matters Farm, Tarrys[390]
- University of British Columbia Farm CSA Program, Vancouver[391]
- Vancouver Island Organic Co-operative, Victoria[392]

Other Farmland Models around the World

When Nathalie returned from the INTO conference in Uganda in September 2013, the way she spoke about farmland protection changed. "The model here in Victoria is pretty hopeful," said Nathalie. "However, not everybody in the world has the same socio-economic structure as we do here in Canada." The INTO conference provided Nathalie with a different perspective. She was allowed a brief glimpse into the workings and management of other models and trusts around the globe, their differing trials and varying goals.

By participating in the INTO conferences, Nathalie has been able to mix with brilliant minds and kindred spirits from all over the world. She has been able to hear others' ideas and share her own. Her passion reminds people of the ripple effect of their actions around the globe.

This global effort to handle the crisis of food security includes: Europe's Common Agricultural Policy, renewed in 2013,[393] the 2013 conference in California hosted by the American Farmland Trust and Napa County Farm Bureau[394] and Vandana Shiva's constant and continuing efforts radiating out from India across the world.

Europe is on the move, with education intentions bubbling up from all sorts of inspired hot spots. The EU Common Agricultural Policy (CAP), launched in 1962, is "a partnership between agriculture and society, between Europe and its farmers. Its main aims are: To improve agricultural productivity, so that consumers have a stable supply of affordable food, and to ensure that EU farmers can make a reasonable living."[395] Of course, as more than half a century has passed since CAP was first implemented, the EU recognizes the need to address the following challenges:

- "Food security – at the global level, food production will have to double in order to feed a world population of 9 billion people in 2050";[396]
- "Climate change and sustainable management of natural resources";[397]

- "Looking after the countryside across the EU and keeping the rural economy alive";[398]
- Providing training for land managers, farmers and farm advisers;
- Providing specialist advice and assistance preparing conservation plans[399]; and
- Paying farmers for the ecological goods and services they provide (cleaning water, providing habitat for wildlife, etc.).

As explained by Stew Hilts, a teacher at the Ontario Agricultural College at the University of Guelph, this last point is "a pillar of the European Common Agricultural Policy and a key component of the US Farm Bill [but] is not yet common in Canada."[400]

In California, a state-wide 2013 conference convened with the intent to "highlight the successes, define the obstacles, and explore new directions for conserving California agricultural land."[401] The California organization, FarmsReach, summarizes the conference in a simple yet sophisticated way, detailing both "What Has Been Working" (such as significant property tax deductions for a farmer's longer term commitment, land trusts, collaboration, public education, creative framing of issues, agencies with vision and authority, as well as constituents who champion all of these) and "What Is Needed" (such as increasing funding, reducing estate taxes and reconsidering farm profit regulations). As Karen Ross, California's secretary of Food and Agriculture, noted, "Everyone who eats has a stake in farmland preservation."[402]

Some of the farmland statistics brought up at the conference are inspiring. For example, California is the world's most productive farming region, feeding 36 million residents, as well as many other US states and countries all over the world. Even though it holds only 2 per cent of US farmland, it somehow churns out 12 per cent of the nation's food value. Despite these amazing numbers, the state of California is not a leader when it comes to land preservation. Its food security, production and systems issue is highly complicated because it involves a population greater than the whole of Canada's. Californians face many of the same fundamental problems as Canadians. For example, despite being big players in agriculture, California farmers lack government funding.

California farmers used to benefit from the Williamson Act, which gave tax deductions to farmers if they agreed to farm their land for a minimum of ten consecutive years. Unfortunately, lack of money has currently put this act into a state of hiatus. California's conservation easements work in similar ways to BC's land trusts. These also cry out for funding.

Focusing on the positives, FarmsReach tells us that most success stories in the state involve hands-on regional and local planners and policy makers. Small counties all over the US are starting community-based, non-profit, independent trusts. Washington, for example, is home to the PCC Farmland Trust, which saved Orting Valley Farms in Pierce County[403] and is a prime example of a trust working in ways similar to, if not more effective than Canadian trusts. The American Farmland Trust is actively pursuing communal, national and global education.[404] If we can learn from history, and one another, we can move more efficiently into a more certain future.

Some of Vandana Shiva's efforts have been covered in other chapters in this book, but we feel compelled to stress that she has inspired much of India and the world to give children a future. Her wisdom presides over a huge number of conversations about food. When the Annam Project – the Good Food Movement – was launched in Kerala, southwest India, in

2008,[405] Vandana Shiva said, "There are only two options – good food or no food."[406] Farmland models worldwide identify with this statement and seek to abolish the myth that chemical agriculture produces more food.

Permaculturalists suggest that creativity is the key not only to good food, but to the infrastructure that supports it – a conscious, communal approach to a holistic lifestyle.[407] Larry Korn, a soil scientist, plant nutritionist and lecturer at the Regenerative Leadership Institute, as well as friend and protégé of natural farmer and philosopher Masanobu Fukuoka, stresses the idea of integrating imagination.

Across the globe, more than 40 countries are joining in the WWOOF (World Wide Opportunities on Organic Farms) movement, which connects real-life organic farmers with younger potential farmers and stirs up excitement surrounding the betterment of food systems.[408] For example, kicking off the year 2014 with positive force, WWOOF Korea branched out to involve itself in a new project, WWOOF CSA. It "builds a mutually supportive relationship between producers and consumers and bridges the gap between urban and rural through organic produce,"[409] extending the reach of the global WWOOFing community into the homes, stomachs and minds of non-farmers or yet-to-be farmers. With an aim to "change the food system and raise awareness of organic farming and sustainable agriculture,"[410] the Korean CSA project is an inspiration that is helping people try out new routes in food production, keep tweaking models that are already in place and continue thinking creatively.

Countless people in every country are fighting against the greed that coils itself so quietly around our souls. The programs in the countries mentioned here are just a few examples of the efforts people are pouring into protecting farmland.

From top: Garry Point, one and a quarter centuries after the Steves family came ashore. This stump is all that remains of a thick forest of trees, bushes and grasses.
Photo: Alexis Birkill, published March 22, 2013, at Straight.com

Christmas Hill, painting by Joanne Thomson (joannethomson.com). Christmas Hill nourished Nathalie's conservation ethic as a child. As she stood on its peak, munching freshly picked wild blackberries, she loved surveying the wildness of Mount Tolmie and PKols (Mt. Douglas), the Salish Sea and, on clear days, the surrounding islands framed by the coastal mountains of both Washington State and the BC mainland.

Farmer Roy Hawes. (Sculpture by Nathan Scott)

Photo: Nathan Scott

193

From top: David and Nathalie Chambers in front of Madrona Farm's organic food stand. Growing together with the lessons provided by sustainable farming, they soon found themselves sharing their experiences with all who stopped by. Their naturally delicious produce inspired an enthusiastic customer base of people who asked why their vegetables and fruit tasted fuller, richer and often sweeter. Logic propelled the next step – taking a stand for saving farmland.

Muddy Lola – whose parents understand the healthy attributes of mud baths, of mucking about in the soil, of seeing for oneself what can grow out of it.

View from Madrona Farm's upper field.

From top: David practises crop rotation at Madrona. According to Harold Steves Jr., he is one of the few who is farming sustainably.

David allows his fields to rest for a season when they need it.

What a treat to see non-GM corn! Madrona's loyal customers wait impatiently for it all year. At the stand, David reveals the secret of the sweetest-tasting corn: pick when milky and eat it raw as soon as possible.

Above, clockwise, from top left: Bee hole; bee on chamomile, bumblebee on yarrow.

Left: snow-laden brassica. Farmers and bees both work through the winter.

Below: Weeding isn't always necessary. Many weeds are friends, and can be nourishing. Many have lessons to teach – about soil and atmospheric vitality, about sharing wealth, about patience, about how to rise above poisons, about helping neighbours versus antagonizing them.

From top: Cover crops blanket Madrona's upper slopes. The trees of PKols mark the border between the farm and the park, behind.

As environmentalists, the Chamberses and their customers have grown proud of the increasing biodiversity at Madrona. Notice all the varying textures.

Look carefully – beyond the upturned boat – to spy one of Madrona's hidden ponds, crucial for the survival of bees. The pond also provides a helpful and hilarious obstacle in Madrona's Chef Survival Challenge events (see page 246 [onwards] for more on these events).

From top: The treehouse, up in the highest corner of the farm, where the Chambers family camped out on the night the frog-seeking teenagers trespassed. (See page 137 for details.)

While David can manage most of the farm by himself throughout the winter months, he provides jobs for two or three hard workers from the community each "summer season," which actually stretches from spring to fall.

From top: Trees are a great example of biodiversity on Madrona Farm.

A glimpse of the trees bordering the division between Madrona and the park on PKols. This forest harbours many memories.

A view of the Blenkinsop Valley Golf Course, from the porch of Madrona's tree house.

199

Clockwise, from top left: Non-native Himalayan blackberries. David and Nathalie are aware of invasive species and work to remove them from their farm. The couple does allow some non-native species to grow along the borders between fields – for example, these Himalayan blackberries, which yield a delicious harvest. Instead of culling every shrub, the Chamberses maintain a moderate number, thoughtfully letting them share the land with other produce.

A sunflower pokes out from its hiding place among the beans. It doesn't belong here, or does it?

A row of the fruit trees that David and Nathalie added to the farm.

200

"We have the capability to destroy other forms
of life; therefore, we have the responsibility
to see that they are not destroyed."

—LEE M. TALBOT, WORLD WILDLIFE FUND INTERNATIONAL[411]

NINE : Identify Vital Farmland

Where and Why?

Twelve minutes' drive from the heart of downtown Victoria,
Madrona Farm rises from the northern end of the Blenkinsop
Valley. Stretching over 11 hectares (27 ac), it graces the southwestern slopes of a 260-metre (853-ft) hill – the remaining
170 hectares (420 ac) of that hill are forested parkland.

Saving Madrona Farm in perpetuity serves many purposes. Its
location in a fertile valley close to British Columbia's capital city
allows it to educate the community about three, major, inter-
twining factors related to sustainable farming. To succeed at
long-term, sustainable agriculture, social, economic and environmental factors must combine. When the first two overlap,
they naturally contribute to cultural stability. When farmers
implement long-term practices with the environment in mind,
the results are lasting and healthy and can initiate profoundly
meaningful societal change.

To maintain that crucial balance between the three factors
that will protect human health for generations to come, Nathalie
and David wanted to make sure that Madrona was permanently

accessible farmland. They encourage existing farmers, prospective farmers and their supportive consumer base to use Madrona's restoration plan as a model by which they might come to respect the ecological integrity of their own land and align their agricultural activities accordingly.

Currently, 6.5 hectares (16 ac) of the farm are being cultivated for mixed market garden vegetable and fruit production. The remaining hectares are forested with Douglas fir and Garry oak ecosystems that embrace four ponds swelling with native frogs.[412] Farming within natural ecosystems restores the land, healing the damage resulting from other human activities. Nathalie and David's goal is to return the land to its sustainably productive base.

In May 2013 a group of Aboriginal and non-Aboriginal people hiked from Mount Douglas Park's base to its summit, where they danced and sang to reclaim its ancient name, PKols (meaning "White Rock/Head"). Erosion-resistant PKols climbed up out of the Vancouver Island volcanoes. The mountain and area, including the rounded rocks in Madrona Farm's current back field, long ago formed a marine bay. When glaciers carved out Blenkinsop Valley, the ocean retreated. Sharp rocks on Madrona's side hill indicate the old banks. As PKols emerged, gradually producing many variations in slopes and soils, marine sediments built up the fertility that characterizes Madrona Farm today. At the southern end of Blenkinsop Valley's string of eight farmstands, Galey Farms' soil yields mollusc shells.

Learning from History and Graduating to a Beautiful Potential

Until now, generous Earth has been forgiving our blind consumption. For decades, however, the indications have become clearer and clearer: we are running out of time.

When David and Nathalie's relationship started blossoming, their passions grew as they studied agroecology, which weaves farming with environmental restoration.[413] This science concentrates on connections in ecosystems, just as many Native peoples have honoured and nurtured connections in their environments over thousands of years. The more they learned, the more questions Nathalie and David started asking: How can we connect food production to earning a fair living? How can we connect socially supportive people in a way that will inspire them to conserve resources? How can we increase production, while sustaining the mineral cycles, which connects to preserving the whole environment?

All biological processes and associated energy transformation are interconnected in complex ways. If we want Earth's food source – and human health – to last and even improve, we must connect at a deeper level to the impacts of technological innovations, policy changes and socio-economic attitudes. If we can understand the science of agroecology, we can heal ecosystems that have been damaged by human activities. Humans are facing either an unmitigated disaster or a beautiful potential.

A Priority Basis for Protecting Farmland

Nathalie breaks down her view of the most important aspects of biodiverse farmland into the following items:

- high ecological and conservation concerns;
- water sources;
- existing species we want to protect, such as species at risk;
- Class 1 to 7 soils, starting with 1 to 3, and restoring 4 to 7;
- high-value agricultural lands;
- cultural usage of the land (following Indigenous traditions of biodiversity);

- historical knowledge (that of early settlers);
- logical climate for growing food, such as that of southern BC;
- a community surrounding it, with stores; and
- most important, a seller who wants to create a legacy.

According to this list, Madrona Farm is situated in an ideal environment. With the help of countless neighbours and friends, David and Nathalie have developed a model with a basic checklist that others are now encouraged to use in the quest for farmland restoration. As their land quality changed, the couple kept monitoring its progress to make sure their long-term sustainability objectives were being met.

They started by initiating agricultural practices that increased biological diversity. For example, Nathalie planted 450 trees in the orchard next to the house. She also maintained the cliff by replanting young trees into a corridor that now stretches from the northern end of the property all the way to the southern end. They have already grown into a tangible band of trees and will eventually become a forest in the middle of the farm. "If we're going to support a dynamic ecosystem, we need to encourage biodiversity," said Nathalie. "Farming without biodiversity is unsustainable." Some farmers farther up the peninsula owned a large part of prime farmland in the Gordon Head neighbourhood on the opposite side of Mount Douglas Park from Madrona. Feeling the financial pinch common to most farmers, they not only lost any sense of biodiversity but also felt compelled to sell their farmland to developers. David and Nathalie decided to lead by example.

"David's dad, a one-third owner of the property, intended to split his share with both his children. This could have financed the back 24 acres, as the front piece, even with less productive soils, was more expensive because of the road frontage. We'd only have had to sell the front four acres, but we turned that idea down," said Nathalie. "Once David and I came together on the dream to protect Madrona Farm forever, we were prepared to walk our talk and see it done, knowing we would die more easily having protected this legacy. Forfeiting our inheritance was a personal and a public statement."

As part of the restoration project, the couple also introduced environmental restoration techniques to manage invasive species, extend corridors, develop habitat and remedy some areas that had been degraded by past agricultural activities. Their invasive species management included getting rid of Scotch broom (*Cytisus scoparius*), common ivy (*Hedera helix*), daphne (*Daphne laureola*), some Himalayan blackberry (*Rubus armeniacus*) bushes, garlic mustard (*Alliaria petiolata*) and non-native bullfrogs (*Lithobates catesbeiana*) that have become a huge threat to the native frogs.[414] These species particular to Madrona threaten biological diversity, and so their removal is essential to the survival of native species.[415]

To ease public accessibility to the vegetables, David and Nathalie enhanced the roadside stand by bringing it closer to the road and remodelling it to include a cover over the lineup area. Thus, even during the rainy months, it would continue its neighbourhood role as an effective gathering place for social interaction and nourishing information.

As David and Nathalie provided both food and education to their customers, their farm was growing economically viable as a business. Part of the culture of the farmstand is a fun, mutual exchange of gossip and information. While David always has some delicious roasted-garlic cooking technique or a caramelized-onion recipe ready to share, many customers are eager to reciprocate with their own creations inspired by Madrona produce. Some loyal customers even bring food samples, such as a famous, annual gift of stuffed zucchini flowers. Other customers bring bowls of plump and tender figs, or other seasonal produce that Madrona has yet to grow. Through these convivial activities, Madrona becomes a valuable member of Victoria's food community.

The stand hosts a bulletin board where the customers line up. While the restoration project was under way, customers could read the latest notice describing the Chamberses' overall goals of achieving environmental, social and economic sustainability. Without each of these three factors in place, the farm cannot contribute to food security.

Madrona Farm is continually on the forward march in the quest for restoration projects. The existing species and species at risk, such as a variety of native bees, red-legged frogs (*Rana aurora*) and green tree frogs (*Hyla cinerea*), are nurtured and the water sources are kept clear of overgrowing and ever-growing invasive species.

Madrona's remarkable variety of soils promises biodiversity success. The agricultural potential of a soil is indicated by its "class." The most arable soil is Class 1 and the least is Class 7. The Agricultural Land Commission describes Class 1 soil as "capable of producing the very widest range of crops. Soil and climate conditions are optimum, resulting in easy management." Class 2 soil follows close behind and is "capable of producing a wide range of crops. Minor restrictions of soil or climate may reduce capability but pose no major difficulties in management."[416] (For more information on soil classification, see the text box in chapter 1.)

The Future of Food

For Nathalie, to be "landlocked" is to live in a world where only wealthy people can afford farmland or good food. "When some investors purchase farmland," says Nathalie, "they often don't intend to farm because they don't understand the true ecological values of the land. This is a major problem. Land protection in the ALR is an illusion. We can't let another piece of land get taken."

Although some non-farmer residents may think it is unfair, tax incentives are given to farmers. Some who buy farmland without intending to farm play the "hay-bale game" to satisfy local tax authorities who want evidence of farm activity before lowering property taxes. Some people who want to escape the city for occasional holiday getaways buy rural farmland and hold it; this is what Nathalie calls "landlocking." Many of these properties have Class 1 and 2 soils and are usually purchased without any type of estate or succession planning. When owners die, their children want to divide it up and often take their cases to the ALR to apply for exclusion from the ALR.

Local governments can significantly influence the provincial ALR's decisions by writing letters of support. Conversely, the public can lobby municipal governments not to support ALR exclusions. A lot of money can be made by developing farmland, as it is often flat and open – easy to subdivide.

Farmlands contain whole or parts of ecosystems with other ecological attributes, such as Garry oak ecosystems, forested areas, watersheds and animal corridors, which need to be considered. These attributes are all essential to sustainable agriculture. Agriculture cannot happen in a bubble: biodiversity, habitat for pollinators, complex soil interactions and carbon sequestering are a few of the items necessary for sustainability. Additionally, big is not always better. Vandana Shiva addresses this in her talks and writings on the future of food. She considers small, biodiverse farms to be meaningful routes to a healthy future.

Metro Vancouver's Delta municipality has some of the best

farmland in Canada, but it's going fast. For example, the port expansion in Delta is adversely affecting one of the four largest bird migration routes in Canada, the Pacific Flyway,[417] which affects falcons, raptors and songbirds.[418] The Delta Farmland and Wildlife Trust encourages Delta farmers to plant winter cover crops through a stewardship program. Because the port expansion devoured food sources for shorebirds, these cover crops are now their main source of food.[419] It's as if Metro Vancouver is saying, "Oh, you want food security? We'll show you food security. We're going to develop our port, pave it all and import the best."

This is all backwards, of course. "Our area here on the southern tip of Vancouver Island, and Vancouver too, has a perfectly logical climate for growing food," says Nathalie. "That's what we need to promote." She claims Richmond's best-known farmland bodyguard, Harold Steves, as one of her mentors.

You first met Harold Steves, Richmond councillor, in chapter 1. He witnessed farmland being sold at the rate of 4,856 hectares (12,000 ac) per year, so he helped set up the first Agricultural Land Reserve in Canada in 1973.[420] Now he owns the last farm on the west dike at Steveston Highway in Richmond, surrounded by 1219-square-metre (4,000-sq ft) houses. "There's this old, museum-like farmhouse filled with gramophones and pictures, with cows grazing on the sedges of the dyke. It's just brilliant," says Nathalie. "I'm working with him, and I feel that it's time to do some work in Vancouver."

Nathalie is on the lookout for farms to protect. Her idea is to start with farms under the highest threat with the most potential for biodiversity. She wants to start on the coast, where the Fraser River Delta provides farmland with ample water and long-term farming potential.

"That land is a community is the basic concept of ecology, but that land is to be loved and respected is an extension of ethics." —ALDO LEOPOLD[421]

TEN : Build Community

Why Do People Matter?

"People," said Nathalie, as she and Sophie were washing and arranging vegetables at the farmstand. "What moves people to conservation?"

"I think everyone is different," said Sophie. "For some people it's a particular statistic that jolts them into wakefulness, and for some it's personal stories."

"For me," said Nathalie, "the conservation ethic started as a tiny seed: a simple awe of nature. I think that's pretty common. But as people grow in years, their experiences of nature grow, too. For some of us – and I include myself in this – the movement toward conservation is the most natural thing in the world."

"Sometimes it's easier to simply farm the fields and enjoy the macro world, living in a biodiverse bubble, forgetting about the destruction going on in the outside world," said Sophie. "But it's impossible to forget for long if you're really connected to the nature you love."

"The experience of finding such valuable, conservation-minded people in my own community really kick-started a hopeful surge upward for me," said Nathalie. "There were so many of them. So, whenever I needed it, there was at least one person ready to feed me hope."

David is one of the deepest inspirational factors for Nathalie. His story touches her heart. As their customers gave wonderful context and meaning to both their lives, David and Nathalie both kept wanting to know more. So, they delved into the history of the farm and the rest of the valley too.

Location, Location, Location

Madrona Farm, the business, is at the centre of politics. The Chamberses helped Elizabeth May, Lana Popham and Dean Murdock in their successful federal, provincial and municipal election campaigns. When people wanted to know which candidates support farmland – they would come to the farmstand and ask the farmers. Nathalie and David began seeing their customers at various protests such as the Spencer Interchange at Bear Mountain.[422] Their farmstand embodied a little box of activism.

The Stand: A Magical Place and a Social Hub

Madrona's farmstand is a place where:

- kindred spirits meet,
- newcomers hear encouragement and support,
- consumers learn about nutritious food,
- gardeners swap growing tips,
- friends share laughter and tears,
- customers tell stories and discuss relationships, and
- comrades debate current events.

The stand is a dynamic place, spilling over with people coming, going, telling their stories, weaving them together, sometimes getting along like peas in a pod and sometimes falling out with each other. It often brims over with love. During the campaign to save the farm, many loyal customers would tell Nathalie, "Of course you can do this!" Nathalie says that they "filled her cup" and that she had no choice but to believe in their faith in her abilities. When she didn't believe it, she had to fake it, especially in front of those who were scared that the Chamberses would lose the farm.

Nathalie admits that she and David are so busy with farming that their customers and apprentices have become their surrogate family. "We're this social hub here, and we love it. That's how communities are supposed to be, with farms as the centre of the social scene, where people can talk about the land."

Because so many of the Chamberses' customers belong to the older generation and have lived in the valley longer than them, they have been able to offer a broader history of the land. Through their stories, David and Nathalie have built a mental map of the neighbourhood. They have found out about all the changes that have taken place over the years, many negatively affecting biodiversity. Members of the Saanich Greenbelt Association took David and Nathalie under their wings, so they could pass on their stewardship of the valley.

Because David and Nathalie had fallen in love with their customers, losing them is always debilitating and devastating. Anita Harwood, now resting in peace, was such a loyal customer at Madrona that she often stopped by to chat, becoming part of everyday life at Madrona Farm. She grew up in the Blenkinsop Valley, in a red house on land where she and her husband Don raised their three children and many chickens. Anita worked at the egg barns for the Galeys nearby, and knew all the neighbours, including David's parents. Over her lifetime, she witnessed the valley's farmland dwindle, and she harboured a lot of local grief.

Anita's heart swelled with joy when she watched David bringing the farm back. She felt a sense of home return, as movement and life grew once again in Madrona's fields. She visited the farmstand every day, and she added David to her mental list of favourites, along with his grandparents and singer Johnny Cash. Saanich purchased the Harwood property when Anita died and added part of it to the park before selling the house on the residential real-estate market.

Ruth Chambers was thrilled to witness the farm's turn-around, too. When David set his hands to the ecological restoration of Madrona, it became a passion project. Seeing land restored to its former capacity has the power to heal the broken bonds between humans and their lost environments. Anita and Ruth were not alone in loving David for supporting the land. Their own stories of loss felt recompensed.

When people happen on the Madrona farmstand, they find a magical place. The developments stretching along Blenkinsop Road pressure traffic to slow down. When people slow down, they actually have time to notice the farmstand. Bicyclists and walkers who take a detour off the nearby Galloping Goose Trail feel like they have found a treasure when they see the stand. Many eagerly return to see what different colourful nutrients they'll be able to pick up the next time.

Word of mouth was the sole fertilizer used to sprout Madrona's customer base, which quickly rose from zero to four thousand. The campaign, through TLC, helped to raise awareness, but political prejudices and the TLC drama compounded competition from the Root Cellar produce store that opened in 2008 less than two kilometres (1 mi) down the road. Customers continued to ebb and flow. The Chamberses realized that customer loyalty to their farmstand went hand in hand with creating awareness and meaningful relationships.

To help people gain an ecological understanding, David or Nathalie would walk customers around the whole farm, pointing out the 450 native trees running north to south through the middle of the property; the 14 new trees that survived out of 38 planted in the extension of their Garry oak ecosystem; the 150 new fruit trees running east to west, which help prevent future erosion of Mount Douglas. They'd discuss their discontinuation of the diesel irrigation pumps, which saved two thousand litres (528 gal) of fuel emissions yearly. Finally, they'd point out the 105 varieties of vegetables in the ground, the three types of grain and the dozens of different flowers. A tour of Madrona Farm was also a tour of local biodiversity, and customers returned to the stand rooted in the richness of nature's complexity, soil stuck to their shoes, frogs singing meaningfully in their memories. As they took a bite of their next, fresh carrot, they understood the importance of the diverse pollinators that helped create that incredible sweetness so particular to Madrona's fresh produce.

Throughout every minute of the campaign, Nathalie felt immersed in a fog of impending disaster, but customers continually brought her hope. David Cutler – a former Edmonton Eskimos football hall-of-famer who eventually became a communications specialist – became a beacon of hope to Nathalie during the campaign. For a while, he'd lived next to Grannie by the farm, and he'd shovelled her driveway. By the time the campaign started, David Cutler had moved and was working at two radio stations: The Q and The Zone. Nathalie contacted him, hoping he'd help her with some strategies.

They met at La Collina Fine European Bakery, attached to the Root Cellar market a couple of minutes' drive from Madrona, and talked about their fascination with the area's First Nations. After an hour, Nathalie felt they hadn't talked about anything. But Cutler felt he knew what he needed to know about Nat.

Her passion for saving farmland had inspired him enough to immerse himself wholeheartedly in the campaign. Nathalie left the meeting with his words ringing in her head: "That's a great idea, kid – and you can do it!" He introduced Nathalie to John Shields, the manager of the radio stations. To her amazement, John generously offered $10,000 worth of free advertising on The Zone and The Q to help her promote the Madrona campaign.

During the ups and downs, Nathalie would call on David Cutler. "Don't drop the ball in the end zone," he'd tell her. "Keep on going!"

Acknowledge The Land Conservancy Friends

Nathalie bonded with many people while working with The Land Conservancy of BC. Bill Turner was a major advocate of education and would jump to offer professional development opportunities to TLC staff. Through a Turner connection, Nathalie found herself at the Land Trust Alliance of British Columbia Conference in April 2011, listening to keynote speaker Story Clark, a professional land conservationist consultant, speaker, author and Wyoming rancher.[423] Clark breaks conservation finance down into methodical pieces that are simple to remember and follow. For example, in her article "Coping with Crisis: Strategies for Helping Land Trusts through the Recession," she reminds us that even seemingly dark situations, such as times of recession, have silver linings: "During hard times, Land Trusts should stay afloat, use networks to secure funding, strengthen relationships, tout the benefits of conservation."[424] Nathalie's methodology already involved strengthening relationships, and affirmation from Story Clark not only provided huge relief but also inspiration.

"Story's words rocked me," said Nathalie. "Conservation without finances is merely conversation, and that became my motto."

During the lunch break, she sidled up to Story with her pad of stapled-together paper and a chewed-up pencil and said, "I'm just sitting by you, writing down everything you say. I want to get it by osmosis. I just want some of you to become some of me." She ended up sharing Madrona's story.

Clark responded: "Whatever you do in your life, tell that story. Tell it to a wide variety of audiences. People need to hear that story."

A couple of months later, when Nathalie was contemplating both TLC's fundraising issues and her desire to keep fundraising beyond Madrona, she discovered that Story Clark was teaching a course that summer at Yale on land trust fundraising techniques. "What?" she thought. "More Story Clark? I have to get in that class. Where do I sign up?" Unfortunately, when Nathalie applied to attend, she received a letter from Clark saying, "I'm really sorry, Nathalie. The class is already full, but maybe next year." However, one week before the class, Nathalie received another e-mail from Clark saying, "You're in!"

Nathalie had a family, a farm and no available funding for such a trip, so attending would be impossible. Sadly, she wrote to Story, admitting she couldn't afford the tuition. By seeming magic, Story offered Nathalie a full bursary. Excitedly, she told Bill the news. He offered more magic, donating all his Air Miles, so her whole trip was covered. The next week – when she stopped in New York, en route to Yale in New Haven, Connecticut – she felt the heat of a city on fire with energy. As people rushed in different directions, their vibrations propelled through her in giant, forward-moving thunderclaps.

Returning to TLC's office in Victoria, her excitement about

Story's inspiring course was contagious. With passion, she shared Story's *A Field Guide to Conservation Finance*. As Terry Tempest Williams wrote in praise of the guide, "This book belongs on the director's desk of every environmental group in the country: in it the fear of raising money is replaced by an understanding of what it means to give. Raising money is always about relationships, our relationship toward the land and each other. Who could imagine a book about money to be a book about love?"[425]

While working hard at TLC, in addition to her work as a pollinator-enhancement specialist and agricultural assistant, Nathalie was continually finding new work to do. As TLC faced financial challenges, Bill continued exploring new avenues. After reading a book called *What Kind of World Do You Want?*[426] by Jim Lord, who works with organizations having fundraising challenges, Bill went to one of Lord's courses in Little Rock, Arkansas. He returned to the office with a pile of brochures and books under his arm. "Please read Jim Lord's book," he implored each of his staff. Having not experienced the author's charisma, few immediately opened it.

However, when Nathalie read it, she felt Jim had written it just for her. She called it her personal "Madrona Farm Protection Counselling Service" because it helped her understand her own feelings, coming down from the high of Madrona's successful campaign. When the daily pressures of the campaign suddenly disappeared, she wasn't sure just what to do with herself. Now, realizing that others experienced the same feelings, she felt empowered. She marched around the office, dropping the book off on co-workers' desks and urging them – as only Nathalie can – to read it. Eventually, one co-worker said, "That's it, Nathalie. You're writing a review for this book." So, Nathalie wrote a review and sent it in to the author.

When TLC's major gifts assistant moved to Vancouver to be closer to her granddaughter, Bill hired Kathy Arnason in her stead. Kathy had taken a Jim Lord class over a decade earlier, and Jim recommended her to Bill. Nathalie wasn't sure how she felt about her new co-worker at first, but it wasn't long before she moved her desk from the quiet back of the office to right up front, beside her. They had their own little space together and called it the "Corner of Belief."

Kathy is one of those people who actively believes in Nathalie's abilities. It takes only one person to trigger a confidence boost in another. Kathy embodies the principles espoused by Lord in *What Kind of World Do You Want?* Working with and being mentored by Kathy made it so much easier for Nathalie to talk about a sustainable future.

"You know what?" Kathy said. "You don't even know how important you are."

What Kind of World Do You Want? envisions having a sustainable future. If we pick up a javelin, we visualize the exact point where it lands – before we toss it. We can create that future in our imaginations, by speaking to it in our minds, by believing in it and talking to other people about it. That's exactly what the Chamberses did with Madrona.

When Nathalie said to David Cutler, "Sure! Of course we can raise $2.7-million," she wondered right away if she were crazy. But people were ready to believe in them. They had their work cut out for them, trying to convince the naysayers and the disbelievers, but it was work that helped change some people for the long term. They convinced some people to entertain themselves with the vision of Madrona's future. The Chamberses asked people to dream, and they all dreamed together.

Later, Nathalie took part in an intensive Jim Lord training retreat in Metchosin, 25 minutes' drive west of Victoria, by the ocean, along with some INTO executives. Thrilled by the experience, Nathalie gained a deeper understanding of why conservation meant so much to her. She learned to reach back to her childhood and think of all the memories of the beautiful places in which she had grown up. She thought of the way these favourite properties had been treated over time, some of them stripped almost entirely of their natural resources or turned into condo developments. What if we could truly protect these places and keep them pristine? Jim Lord, and the retreat, helped Nathalie to define her conservation ethic. She describes it as the expression of her love of nature, through action.

Nathalie also explored the Appreciative Inquiry (AI) technique, which "is about the co-evolutionary search for the best in people, their organizations, and the relevant world around them."[427] Appreciative Inquiry is an interview and listening technique; after listening carefully, acutely to another person speak, you are able to ask questions of consequence. Through her studies, Nathalie felt she could reach people who thought very differently from herself. Nathalie's new frontier lay beyond the choir. By applying Appreciative Inquiry, she knew she would be able to open people up to talking about farmlands and biodiversity. AI turns the key, allowing people to unlock their hearts for answers to questions.

Among other people she met during some of the training conferences that INTO offered, Nathalie became good friends with Rob Macklin, who is the head of food and farming at the UK's National Trust, and chair of the judging panel for its ongoing Fine Farm Produce Awards.[428] In April 2012 she flew to the UK to help Rob with some of his work.

When Rob and Nathalie drove to Wales together, they shared dreams and talked about the difficulties they'd encountered while implementing the agroecological agenda. Nathalie described the pollinator program, which they didn't have at the National Trust. She used Appreciative Inquiry during their discussions.

Nathalie understood and discussed the tricky balancing act between providing for a family and the risky business of making honest but revolutionary statements and decisions in the workplace. Because of their connection, Nathalie and Rob felt that by combining his event, the Fine Farm Produce Awards, and her Chef Survival Challenge – both which are similar in nature – they could create a pretty huge, amazing event. "If we can't do what we want at this level," she said to Rob, "we can go to the next level, up to the top level, and funds will rain down on the trusts around the world."

Acknowledge the Chefs

Chefs at delicious eateries all over Victoria eagerly order fresh produce from Madrona, or support Madrona in other ways. Their loyalty to the farm and their passion to provide local, organic food to the community drives many of them to participate in Nathalie's Chef Survival Challenge and broadcast the sources of their ingredients in their creations. In the notes section of this book, we provide a list of these creative chefs,[429] both for readers who live in, or travel to, Victoria, and for others who may wish to learn from referenced articles and interviews about them.

Acknowledge the Donors

The fundraising adventure was a game of suspense, with every donor hoped for but not expected.

When Helen and Glen Sawyer, local Victoria philanthropists and lifelong butterfly collectors, called up the Royal BC Museum to inquire about donating money to butterfly preservation, they were passed along to James Miskelly, who worked on Taylor's checkerspot butterflies (*Euphydryas editha taylori*), now extinct in the Garry oak ecosystem. The elderly couple ended up donating $150,000 to the butterflies. James also turned them on to Madrona, having been to a presentation at the University of Victoria the night before, where Nathalie spoke about the farm's 1.4-hectare (3.5-ac) Garry oak ecosystem. Nathalie met and fell in love with Glen and Helen. She was inspired when Helen told her that she believed in three different destinations for money: family, conservation and saving a penny for a rainy day.

When David and Nathalie led the Sawyers on a tour of Madrona Farm, David had to carry Helen through the passageway into the living room of the farmhouse because she had a lot of difficulty walking. As he lifted her up, he said to Glen, "I hope you don't mind me carrying your bride over the threshold!"

Jeff Campbell, an advisory lawyer for Madrona, read an article in *The Globe and Mail* about the Sawyers' lifelong careful planning and savings, their generous donations to local hospice care, to the art gallery and to Madrona.[430] It inspired him to help Madrona find the rest of its needed funds (see chapter 11), and from there, the campaign galloped down the homestretch. After hearing Jeff's expert and confidently optimistic opinion, Nathalie and David finally took their first holiday in years.

People poured in to help from places that Nathalie would never have expected. At first, the Woodwynn Farms campaign competed with Madrona for attention and donations from the community. However, when Woodwynn Farms transferred its own $400,000 Farmlands Trust donations over to Madrona (as explained in chapter 11), both former and current chairs of the Farmlands Trust, Barbara Souther and Ed Johnson, also became enthusiastic advocates for the Madrona campaign.

Likewise, the success of Ed Johnson's challenge, which evolved from him flippantly saying he would match a $300,000

"Jimmy Pattison" donation to a serious offer matching anyone's contribution of $200,000, which he posted as an official challenge on YouTube.[431] Guy Dauncey also spread the word in his popular *EcoNews.*[432]

Before a month was up, and seemingly out of nowhere, Mel McDonald, who feels angels are always sending him messages, received a declaration that he should match Ed Johnson's offer. Such donations, arriving just as they were most needed, made the campaign impossible to quit. Through such deeply touching meetings with so many caring people in her community, Nathalie found that she couldn't sink.

Create Ongoing Opportunities for Community Engagement

Nathalie and David feel that they are rich in community, and produce! And they love to share. All over Victoria, grassroots movements are growing into beautiful organisms of change. Madrona donates its leftover produce each day to Food Not Bombs, which shares a free vegan meal every Sunday at 3:30 pm beneath the tree at the corner of Vancouver and Pandora Streets.[433] David and Nathalie also used to drive downtown daily in order to drop off food at the Upper Room, the Mustard Seed and the Victoria Cool Aid Society.

Lifecycles Project Society is an organization based in Victoria, whose famous Fruit Project reduces waste all over town and provides free food for those in need.[434] Lifecycles volunteers come together in their cheery, common passion for fruit and pluck to their hearts' content, all the while mapping the ins and outs of Victoria's unused fruit trees on private properties. Some of the organization's luscious proceeds go to the Mustard Seed, a local non-profit dedicated to the fight against hunger.[435] Madrona has donated fundraising proceeds to the Lifecycles umbrella organization, which also heads up initiatives[436] such as these:

- Grow a Row,[437] inspiring people to grow an extra row of produce in their backyard gardens, to be distributed to local programs such as Our Place;[438]
- Growing Schools,[439] bringing together young students and volunteers to deliver education programs about urban ecosystems through the creation and exploration of school gardens;
- The Pollinator Project,[440] creating urban pollinator corridors;
- Sharing Backyards,[441] connecting people who have space for growing food with people who want to grow food but don't have space; and
- The Urban Agriculture Hub[442] and Urban Agriculture Program,[443] providing education, support and networking opportunities for biodiversity projects – such as garden creation, urban agriculture and food preservation – in the form of an online resource; and

At the start of 2011, Garden Works,[444] a nursery within three minutes' walk of Madrona, partnered with TLC to create a Giving Tree program. They picked up live, potted Christmas trees from peoples' homes around Victoria – Douglas firs, Fraser firs and Colorado spruces – and transplanted them to Madrona Farm.[445]

The more we look around, the more we see, springing up on urban streets, grassroots movements that yearn for a fuller sense of regeneration and community. As people are putting their ideas into action every day, the list of grassroots movements is always growing. These are some in Victoria alone:

- Eat Local Grown, "a crowd-sourced community driven tool that helps you find, rate and share locally grown food. Its categories include farms, farmers' markets, restaurants and grocery/co-ops, and all kinds of artisans like butchers and cheesemakers."[446]

- FarmFolk CityFolk (FFCF),[447] a local, "not for profit society working to cultivate a local, sustainable food system. Our projects provide access to and protection of farmland; support local growers and producers; and engage communities in the celebration of local food."[448] Their famous annual fundraising event, the Feast of Fields,[449] is a four-hour wandering harvest festival, hosted by a different farm each year. The Madrona campaign teamed up with the Feast of Fields to garner publicity and support. While hosting other events, such as a Permaculture Design Certificate course, FFCF also has a brilliant variety of farm, farm-city and education projects under way. Some of these include the Micro-Loan Fund, Seed Security, Young Agrarians, Get Local and Growing Green Law Reform.[450]

- Island Chefs' Collaborative,[451] a group of 50 (and growing) chefs on Vancouver Island who are focused on sustainable food and agriculture.

- Transition Victoria,[452] a regional initiative that will "joyfully raise awareness, nurture personal transformation, build social connections, support visions and actions to help create more resilient, democratic, vibrant communities that no longer need to rely on fossil fuels as their primary energy sources."[453] Some of the group's current projects include The Linen Project,[454] Transition Streets,[455] the Capital Nut Project,[456] Springridge Common[457] and Home Groups.[458]

- Victoria Forest Action Network (VIC FAN)[459] creatively and actively supports permaculture principles and protests pipelines in the Greater Victoria Region, among many other forward-thinking endeavours.

Acknowledge the Community's Strength and Expand

Within our communities, Nathalie encourages us to:

- work together as a force;
- employ the "Yes, we can!" style of storytelling;
- tell stories: at the farmstand, in the media, at events; and
- involve people in the cause by building a volunteer base.

Building new connections within an already subsisting community is a start, but why should there be limits to how big we grow? Nathalie dreams of moving beyond the local and joining the global community. For example, she dreamed of having Prince Charles and the Queen come to Madrona and help with the campaign. She imagined the prince whooshing overhead, along the AdrenaLINE zipline at one of the Chef Survival Challenge events. At one point, she even commissioned Sean Ryan, her friend and the artist for Phillips Brewery labels, to sketch a picture of the Queen astride a Massey Ferguson tractor.

After each of her excursions, to the UK, Uganda, the United States, the Canadian prairies, eastern Canada and the nearby Gulf Islands, Nathalie returns home with an expanded perspective and an expanded sense of purpose. She maximizes the value of her expeditions by mindfully gathering information wherever she goes, talking to the people and integrating what she learns into her life's work.

If we can all remind ourselves to tell our own honest stories and listen to the stories that are being told every day, we can move toward a community that is more like a body, with feet and hands in outposts all over the world.

"Fundraising is the gentle art of teaching
the joy of giving." —HANK ROSSO[460]

ELEVEN : Fundraise, Fundraise, Fundraise

Sourcing $2.7-million

Some people donated less than $20 to the Madrona campaign,
while others contributed tens or even hundreds of thousands.
Some, unable to donate finances, gave generously of their
time. Others did both. Eventually, the many different sizes and
shapes of donations juggled together to complete the puzzle.
Quite simply, the $2.7-million shook itself out of the commu-
nity when people, one by one, understood the goal.

On the final day of the campaign, The Land Conservancy
of BC released an overarching statement of closure, published
by Heritage Canada, The National Trust: "After a twenty-four
month campaign and close to 3,000 donors, The Land Conservancy was proud to announce Madrona Farm was saved and
would remain in agricultural production forever. Thanks to
overwhelming public support as well as major contributions
from the Farmlands Trust, Victoria Foundation and local

Victoria residents Ed Johnson, Mel McDonald, and Helen and Glenn Sawyer, TLC is pleased to add Madrona to its list of 'special places' in B.C."[461]

As mentioned earlier, the Farmlands Trust (FLT) chairman Ed Johnson announced his desire to match anyone's donation of $200,000 in order to bring individual donations to a total of $500,000.[462] Before long, Mel McDonald, a Victoria resident, made his offer. "The Chamberses are really hard working people and pretty selfless," Johnson said. "They have given up their own inheritance from David's father and at the end of the day all they'll get is a chance to lease it and walk away with no equity. I really respect people like that."[463]

Meanwhile, Helen and Glenn Sawyer also contributed. Over the years, the couple has contributed more than $2.5-million to charities. Helen has supported the Victoria Hospice Society and the Art Gallery of Greater Victoria, and when she saw the Madrona campaign emerging, her passion rose again, to the tune of $80,000.[464] Helen learned about giving from a very young age when her father encouraged her to save pennies for church each week. Over time, her focus shifted from one worthy cause to the next, but she has consistently given thousands away. Of Madrona, she says firmly, "This property should never go to development."[465]

Not every individual donation story reflects such a level of philanthropic character or care for the community and environment. Tony Young's donation demonstrates a very different kind of giving. He and co-developer Jim Duncan, both from Victoria, faced corruption-related charges in 2009 to do with removing land from the ALR while working on a new housing development for Sunriver Estates in Sooke.[466] The Supreme Court of BC decided that a kind of compensation needed to be paid to make amends and show goodwill to farmland and the community. Thus, Young and his lawyer, Jeff Campbell, chose to donate to the Madrona Farm campaign. The money arrived in March 2010, just in time to finish off the campaign. This saving gift, and others, helped the Madrona team learn to maintain faith and keep pushing forward in the midst of uncertainty. Hope needed to keep its flag raised, right until the final moments of the campaign.

Forfeiting Inheritance

Without education efforts pouring into the Greater Victoria community throughout the Madrona campaign, we'd have had no reason for hope. While the team was actively fundraising for about two years, education – the primer before the paint – started before the first dollar's appearance. While some minds needed enlightening, others just needed reminding.

Philosopher, environmental activist and writer Vandana Shiva said, "You are not Atlas carrying the world on your shoulder. It is good to remember that the planet is carrying you."[467] All beings brought onto Earth through a biological process are connected to nature. Some of us simply need to dig a little deeper to unearth that connection.

Crucially, the public needed to start out by understanding that David and Nathalie were not raising the money for themselves or for *their* farm. When the couple first met, David owned a share of Madrona as part of his inheritance. However, Nathalie never wanted to own the land.

There is a distinct difference between the responsibilities of stewardship and those of ownership. Stewardship involves providing care through dedicating time, while ownership is about taking control by dedicating money. As a famous Native American proverb reminds us, "We do not inherit the land (or our future) from our ancestors; we borrow it from our children."

Ownership separates us, whereas stewardship brings us together. A steward sees through the illusion that our lives don't affect others, because she is always mindful of those who come after her.

Responsibility. Caring. Protection. These words evoke different meanings and varying standards for each person who reads them. But we all need to pay at least some attention to protecting our health, sustenance and survival on a global scale. The ALR represents a start toward protecting land, but we need to augment it as we march into the future. Some people who have heard of the ALR wonder why so much land still needs saving, but in reality it protects less than 5 per cent of BC land, and even then the ALR often allows farmlands to be deleted from the reserve or used for non-agricultural, non-biodiverse purposes.

Madrona supporters would like to see the ALR add tenets for organic methods of farming. Currently, it allows a very broad range of farming practices, from uses such as passive recreation to scenery viewing. Biodiversity conservation is only one "accepted" use.[468] More than merely "accept" it, we would do well to actively encourage biodiversity within the ALR. The ALR is a starting point, but we must aim for biodiversity to strengthen agriculture nationwide, thereby strengthening food security.[469]

At the start of the Madrona campaign, a lot of people didn't understand the situation or the reasoning behind the fundraising. The Chamberses fielded incredulous questions: Why would you forfeit your inheritance? Where are you going to live when you retire?

David and Nathalie had already thought deeply about these highly personal questions. In reality, though, Nathalie owns another piece of land not far from Madrona, one that she's been renting out for years. Avoiding a rash decision, they instead inspired the community by donating their own inheritance from the sale of the farm, for future generations in perpetuity.

This donation inspired a lot of local news coverage. At the same time, because of conflicting facts, people still struggled over what to believe as the campaign went on. In a country with local, national and global news so readily available, we still find ourselves ignorant of certain perspectives. Immersed in our daily grind, we fall behind, overwhelmed by all the information flying at us from all angles. If we take a moment to look up and be aware, we'll realize that we do always have a choice: to know or not to know. Let's not be willfully ignorant. Instead, let's prioritize, read well-researched news and choose to make ourselves aware. And when we are overcome with uncertainty, why not go to the source? Madrona's location was central, and David and Nathalie were ready and willing to provide clarification.

Dreams Coming True

While enlightening others, Nathalie and David were continually educating themselves. Nathalie feels strongly that she emerged from the campaign a new person. After recognizing her global grief, she labelled herself a "deep ecologist." Rather than wallowing, she embraced hope, found concrete ways to keep her positive energy flowing and moved forward.

Shortly before the farm hosted its first Chef Survival Challenge in 2008, David and Nathalie went on a road trip across Oregon, Washington, California and Arizona. The design of the Obama campaign grabbed Nathalie's attention. She returned home thinking "Yes, we can!" about Madrona's campaign. Repeating this phrase reinforced her own mantra of hope each day.

The bees provided another source of connection for Nathalie. When the numbers and naysayers, phone calls and

Smart Car trips all over town became all too heavy a current against her, she fled to the fields. The cover crops were a sanctuary more than once, always ready to embrace her. Lying down, she found it possible to be transported, hypnotized and mesmerized.

Just as Nathalie felt the middle-C sound of the buzzing bees alter her state of mind, she consciously became more attuned to the different vibrations of community. In particular, she noticed when people were taking on her cause with their whole hearts, and fitting themselves in as an extension of it. Nathalie knew she had reached a tipping point when her efforts touched people in such a way that her stream became their collective river.

At one point, a friend who worked at the Capital Regional District water services department became involved by supplying the Chef Survival Challenge with filtered water. Her outpouring of energy affected her daughter, who held a lemonade stand with the proceeds going toward protecting Madrona.

Before the first Chef Survival Challenge, mini, one-off fundraisers included donations of birthday money, lemonade stands, strategically located coin jars and dollar campaigns. Others reached out to a broader group, such as the Phillips Brewery's Blackberry Hefeweisen, a beer made from Madrona's wild blackberry harvest.

While all these small campaigns succeeded in their own ways, Nathalie felt the community was ready for a bigger event in which everyone could participate. What started out as a topic of morning coffee conversations was about to turn into a flurry of activism. TLC's agricultural liaison, Ramona Scott, hired a woman named Christina Torgenson (who later became involved with the Islands Trust), and together they initiated a brainstorming session where people voiced their fundraising event ideas. Spurred on by the continual ALR deletions,

Nathalie and her TLC co-workers produced many enthusiastic suggestions that quickly turned into a wall of lists. The lists even included asking a couple of famous people to encourage participation.

The brainstorming sessions gave birth to the first Chef Survival Challenge. Prince Charles's continuing public statements on the issue of global food security and sustainability inspired Nathalie's efforts.[470] Acting on her fantasy of having HRH at the event, she wrote to him and received a letter in reply, passing on his best wishes for the event. Nathalie felt delighted with the positive energy flowing both to her and to the Victoria community.

Crucial Ingredients

One ingredient needed to fundraise successfully is gall. Or nerve. Or guts. Or whatever vital energy or organ it takes to drive people to overcome their fear of wild success. Remember going door to door with a childhood sports teammate to raise money for new track suits? At least back then we were selling chocolate or soap and people couldn't resist our cute, rosy faces.

Nathalie and the team were trying to "sell" people something less tangible than chocolate or soap, but so much more real and necessary. She was pitching a vision and an ideal.

Some people, like Nathalie, seem made for canvassing and even feel high from it. The words flow generously and with ease. The world can't leave the talking up to these nervy people alone, though. If we're looking for true, heartfelt success that breeds more success, we all need to contribute. With many roles to play, everybody can donate.

Another necessary ingredient in a fundraiser, of course, is passion. Passion and bravery.

David and Nathalie weren't afraid to speak up, to protest

pollution and deceit and unsound planning. They were particularly protective of the Blenkinsop Valley surrounding their home. For example, the Blenkinsop Valley Golf Centre, just a couple of properties away, radiated light pollution in the area, so David and Nathalie protested fiercely.

At night, Madrona's fields were like a Walmart parking lot, totally lit up. Nathalie worried about the owl population in the area, knowing that light at night was unhealthy for them, as well as for herself and her family. "We need moonlight at night, thank you very much!"

Before launching into an outright protest, Nathalie talked to her neighbour, Dr. McPherson. He'd been protesting the lights for over ten years. According to him, the lights were illegal. In the last century, Ruth Chambers and her friends had carved out time to make and enforce the Saanich bylaws for the golf course, which required all lighting to be shut down at night. Keeping the lights on overnight signified a broken covenant. At a special committee meeting in 2008, David Chambers mentioned that the light restrictions on the golf course had not been enforced since 2000. Many people were outraged to hear this and started to side with David and Nathalie.[471] The immense bright lights were installed without permits, but because a covenant is only as good as the people who enforce it, the driving range with its mini-golf course had managed to circumvent it for all those years.

Nathalie's affinity for her deceased grandmother-in-law drove her to take on the cause. In return for her complaint to council, she was informed that the golf course would be given 90 days to make the change. She was nonplussed.

It was no surprise that after 90 days, the lighting schedule remained the same. Nathalie followed up with another complaint, and this time, feeling she needed to work directly with her neighbours instead of relying on a middle man, she wrote a letter to the Blenkinsop Valley Community Association (BVCA). While pushing her case seemed way out of her comfort zone, she knew she was ethically seeking a legal right. She didn't foresee the ambush that awaited her. When at the BVCA meeting, almost a dozen people berated her at once, Nathalie felt completely abused.

Terrified, she cried for three days afterward. Afraid to walk home alone at night, knowing how intensely both developers and golfers disliked her, she realized her unpopularity didn't bode well for the start of Madrona's campaign. At the same time, she was becoming a recognized entity in the community. She began contemplating the validity of the adage that any publicity is good publicity.

At one point during the protesting frenzy, David brought a lamp to the meeting and shone it in a representative speaker's face to demonstrate the awful essence of light disturbance. People were impressed, amused and indignant. The couple was definitely making a memorable mark on their community. In the end, their protesting of the golf course lights won over some change. The course lost its permit for the lights, and after all the trauma, David and Nathalie allowed themselves to feel a little smug over the victory.

While Nathalie's innate driving force means she vehemently protests when she senses injustice, David needed more of a push, because he would rather stay out of the conflict. At one of the local council meetings, Nathalie fought especially hard to get David involved. She tried to impress on him how much the community respected him. She acknowledged his power, rather than trying to empower him. "People can't empower other people," she says. "They're already empowered." She felt her role was to encourage and respect David's role.

"Farmers are like doctors," she says. "Having a relationship with farmers is very important. She felt it crucial to motivate David to go to these protests. People would see him and recognize him. They'd make the connection between the image of him and the idea of food."

When David does decide to go for it, he really goes for it. Nathalie remembers falling more in love with him after seeing him "get all political" and involved in protesting at the council meetings. One time he brought what Nathalie calls his "tickle trunk boxes," the blue bins that usually transfer his produce from field to truck to stand or restaurant. He started to have fun with protesting and holding everyone in suspense with his mysterious boxes before making the big revelation. He pulled out two watermelon and two zucchini plants – one male and one female of each. Many people weren't even aware that it takes both a male and female plant to reproduce these most delicious and favourite of summer fruits. David taught them the difference, and of the need for pollination. He made the point that the more Blenkinsop Valley allowed development and suburbanization, the more they'd be squeezing out the bees, and thus their favourite food.

Saanich stepped up to the plate in some ways, with a local area plan and urban containment boundaries both working toward protection of the Blenkinsop Valley.[472] However, more needs to be done, because realistically most of the land is zoned as rural-residential. Only 8.1 hectares (20 ac) north of Madrona Farm's 9.7 hectares (24 ac), and Galey Farms' 24.3-hectare (60-ac) flood plain south along Blenkinsop Road are zoned "agriculture." The drive to develop is still strong, as more and more land is taken from the ALR. In spring 2014, changes were made to the ALR that will allow more agribusiness (such as food processing) for farmers in Zone 2, whose growing seasons are

234

shorter than those in Zone 1.[473] (Zone 1 includes BC's Island Panel Region, the Okanagan Panel Region and the South Coast Panel Region; and Zone 2 includes all areas of BC not in Zone 1.) Some people argue that any land-use changes within the ALR weaken it and thus believe that the land will become more susceptible to being removed.[474] The Saanich Greenbelters passed on the farmlands preservation torch to David, Nathalie and their farming neighbours. Nathalie takes it very seriously, and is holding on to her dream of protecting the whole Blenkinsop Valley.

Nathalie's passion extends to creativity. She commemorates their War of the Lights with golf clubs bent and woven into her driveway gate. At first, when she gave farm tours, she suggested that people donate golf clubs to add to her gate collection. However, a lot of her donors are golfers, so she needed to recognize the dangers of generalizing.

After all the protesting, Nathalie and David were getting comfortable with their newfound, on-the-offensive mood. When the previous owners of Beckwith Farm, just down the road, began using propane cannons to scare the birds away from their blueberry crops in the summer of 2010,[475] the Chamberses rallied in protest once again.[476]

David and Nathalie's protest list grew as they faced clashes with their conservation ethic. Recognizing problems meant finding solutions. Anticipating conflicts provided the key to their preparation process. They started the campaign knowing themselves well enough to believe in the strength of their own voices.

Adding your voice to a crowd or a choir brings up the volume, but it's hard to measure the worth of the chorus if everyone is singing in unison. If one singer starts a harmony, though, he is immediately in the spotlight. Imagine, then, if a singer decides to step out in discord. Oh, dear, how noticeable! David and Nathalie became infamous for their dissonance. They were rebels with a cause. They were yelled at during meetings, but only for speaking from their hearts. The friction grew more than uncomfortable. After some confrontations, Nathalie admitted to feeling sick for weeks.

Acceptance of discomfort: yet another helpful ingredient for fundraising.

The rescue of Madrona Farm was a lofty goal. Accepting the challenge, while a huge and difficult decision in itself, instigated a greater struggle, with confrontation at every bend. The farm is protected today because the community was willing to face confrontation and embrace the challenge to raise $2.7-million.

Another bump in the path emerged in the shape of TLC having to obtain a mortgage to bridge a pending payment promised by Saanich Municipality. In 2010 Saanich promised to pay $50,000 per 0.4 hectares (1 ac) of land to TLC as part of the purchase arrangements – in order to add a small portion (1.41 hectares/3.5 ac) of the upper slopes of Madrona to Mount Douglas Park. They could not/would not pay that money until the parkland could be subdivided from the farm. TLC's small mortgage was finally paid out when the parkland was clear. Saanich paid its share in May of 2011.[477] Although the route the campaign took seemed circuitous and at times fractious, such journeys provide reminders of the flexibility and compassion we all need to solve problems together. We, the people in a community, can discover in ourselves the means to achieve greatness. If we can work through the hurt, make unconscious actions conscious, help each other see beyond a previous lack of foresight – in short, gently educate each other – we will succeed. The process of fundraising is about people, either working in conjunction as a community or working through disjunction to become community.

The ability to solve problems and work through challenges is integral to any campaign. Patience is mandatory. To cover a worst-case scenario – say if The Land Conservancy faced bankruptcy or had to start divesting properties – the Chamberses made sure that Madrona Farm would go to another organization, such as Heritage Canada, The National Trust, with similar protection values. They also made sure the achieved lease restricts new leaseholders to using/caring for the land only in the same way the Chamberses do: to supply the local community with organic produce by supporting biodiversity, thus working toward the goal of national food security. The Land Conservancy, and the community, now owns the land and personal ownership can never be regained. The challenges – price changes, obscure budgeting and even possible bankruptcy – resulted in a real-life model that promises to be sustainable over time.

One of the positive aspects of the Madrona Farm campaign was the non-competitive approach various media took when covering the story. They each delivered an all-encompassing support campaign, explaining to the public that putting Madrona into a land trust preserved it for the common good. As everyone worked together, Nathalie grew more and more inspired. This nourishing aspect of community involvement became the most crucial ingredient to fundraising.

Getting Traction

When Nathalie first formed the idea to protect Madrona Farm, she talked about it with David for a year before bringing it to the community. As customers arrived at their farmstand at the end of the farm's long, gravel driveway, she began talking about her views on stewardship versus ownership, and the huge risk of the farm falling to other uses once she and David moved on from it. Unable to hold her ideas in, feeling pressed to share, her drive to educate people grew naturally. Gradually, she gathered together a board of supporters with the help of TLC.

The Friends of Madrona Farm Society was formed in March 2008, made up of anyone who took Nathalie seriously enough to join and attend monthly meetings. A couple of months later, they unofficially launched the society at a sustainable feast put on by Food Roots and Share Organics in the Oak Room at the Fairfield Community Hall. During May's meeting, Karen Platt wrote and presented a beautiful story about David and Nathalie's plans.

Then-BC premier Gordon Campbell's Climate Action Dividend rebates spurred on the first big fundraiser that really started the campaign rolling. Through YouTube, the Victoria *Times Colonist*, CHEK News and other media, David and Nathalie kept awareness of their cause on the rise. They suggested that people re-gift their $100 Climate Action Dividend cheques, issued by the provincial government, to Madrona's cause, and this suggestion brought in the first, overwhelming $100,000 in June 2008.[478]

When the price of the farm was renegotiated in June 2008, the inconsistency caused some doubt among donors, and so the pressure was on Nathalie to prove herself trustworthy.[479] After a binding agreement with new numbers was finally signed in July 2008, some of the uncertainty settled down.

At each important point in the campaign, the media stepped in eagerly, encouraging a natural progression of raising awareness. Shaw TV in particular proved to be a ready and willing promoter. However, at times, different media pieces conflicted with one another, causing confusion in the community. Because of this, Nathalie and David tried to be even more physically and personally present with people, always ready to answer their questions.

Cooking demonstrations also succeeded in educating and raising awareness. David and Nathalie started their cooking shows in 2006[480] and held them intermittently throughout the campaign. David's background as a camp cook in the Yukon, and as a traveller influenced by fresh food throughout Japan, helped him remind everyone of the simplicity of making a delicious dish at home, with the right ingredients. Customers' anticipation of the weekly event meant both a business booster for their farmstand and a chance for people to ask questions.

A media kick-off with Grania Litwin, a Victoria *Times Colonist* editor, preceded the first Chef Survival Challenge in October 2008. She wrote a really sweet story about Madrona,[481] and all of a sudden, the parking lot was packed. Ladies brought their lawn chairs and flirted with David, while he and Nathalie pumped away at the food, selling it all by the end of each day and giving away free cooking tips just as fast. The mouth-watering, "Madronafied" recipes they handed out provided even more connections between farmer, customer and food.

Juggling Fundraisers

Many promising ideas hung in the air after the brainstorming during that initial session with Christina Torgenson. Instead of focusing on one, Nathalie decided she wanted to reach as many people as possible by following all the best ideas. In August 2008 the Blackberry Blitz reminded people about where their food comes from as they volunteered to stain their hands purple. Once the volunteers had picked and juiced all Madrona's blackberries, Phillips Brewery mixed in the beer, releasing it in September as Blackberry Hefeweisen.[482]

A month after the berry gathering, Madrona participated in the annual Feast of Fields event, in which gourmet food lovers in the Lower Mainland, the Okanagan and on Vancouver Island experience some of the province's best food offerings.[483] As well as supporting FarmFolk CityFolk's fundraiser for agricultural initiatives and a more sustainable and food secure future, taking part in the event was a practical step in the scheme of raising awareness about the farm. They were excited to contribute to FFCF's Local Food Micro-Loan Fund.[484] Not long after this, on October 4, 2008, Nathalie launched the first Chef Survival Challenge.

That Christmas, Nathalie suggested people give a gift of long-term food security in the region by donating to Madrona's campaign. As David advised, "Instead of people donating a goat in Africa or planting something in New Guinea, we are suggesting this Christmas they give a gift of long-term food security in this region." The local newspapers broadcast the fact that 25 businesses were donating in different ways, including clothing retailer Rebel Rebel holding a bazaar, Fix Hari Garage having a fashion show, and La Collina Restaurant baking tractor cookies – with partial proceeds going to the farm. The consciousness of the community had grown to such a level that Christmas gifts raised another $5,500. By the end of December 2008, the campaign had reached its first goal of $250,000.

Continual rebalancing and fluidity was needed for the campaign to stay alive. Within weeks of the New Year starting, Nathalie organized the Rock and Royal Tea Party[485] at the Blethering Place Tea Room and Restaurant in Oak Bay. People bought the $40 tickets and enjoyed Madrona-fresh produce transformed into a three-course meal, with the event catered to those who love good food, good music and good poetry. At the fundraiser, Nathalie spoke up about the ongoing efforts to save Madrona. The Chamberses realized that consistent momentum formed a mandatory ingredient to keep their campaign fresh.

In the next few months, while Nathalie kept busy

planning each next step, as well as the second annual Chef Survival Challenge, David uploaded a YouTube video called "Save Madrona Farm with $1."[486] This mini-campaign, while a branch of the larger structure, also spread the responsibility, reminding everyone that "every single dollar [we] spend in this world can make a difference." Because this micro-fundraiser allowed so many people to be a part of the project, the sense of stewardship over Madrona expanded to include many contributors. The farm was truly becoming a part of the community. All over Greater Victoria, groups of people started their own initiatives to raise money for the Madrona campaign. La Piola Italian Restaurant, Camille's Restaurant, Caffè Fantastico and La Collina Bakery, as well as University of Victoria's Environmental Studies Student Association, all contributed.

In March of that same spring, Woodwynn Farms was finally sold to Richard Leblanc of the Creating Homefulness Society.[487] By this time, Woodwynn's rivalry with Madrona had turned into support. With TLC going through leadership struggles, and both the media and customers taking sides, Nathalie became determined to refocus public attention by advertising her farm tours more widely. Although the tours did elicit donations, Nathalie focused on their educational value. As she walked with people over the earth their food grows in, and while looking out across vistas of beautiful but unprotected land, she reminded them of the need to defend their own backyards.

Farming with biodiversity takes people, bugs, birds, animals, trees and plants back to their roots. Madrona is continually increasing its biodiversity, and when people take tours of the farm, or party at the Chef Survival Challenge, they relax in the fullness of the environment. At that point, they naturally understand its value compared to a vast, monoculture, industrial farm. By July 2009 root reminders like these led the community to donate enough for the campaign to reach its second goal of $750,000.

The next deadline was January 31, 2010, and the second Chef Survival Challenge in October 2009 raised a total of $17,000. A powerful bond had grown between people as they acted together on their common interests. They hugged and cried. Moments like those motivated each next step.

As the Christmas season of 2009 started to break over the city, David and Nathalie created another media campaign asking people to give the gift of food security. As they explained, "This is a gift that keeps on giving."[488]

Nathalie created another fundraising effort she dubbed the MOJO on LOJO Business Blitz. She created sticky labels emblazoned with a red tractor and the words "Protect Madrona Farm Forever." Applying them to mason jars, she cut coin slits in the lids and put the jars on the sales counters of every store on lower Johnson Street in the heart of downtown Victoria. The fundraiser succeeded by encouraging people to take on the cause personally. Seemingly small fundraisers like this one helped the campaign reach a tipping point, at which people began to invest themselves in fundraising, earnestly and out of like-mindedness, building community gradually until together they created financial success.

While the community stepped up in many ways, the campaign still required an extension. Its January 2010 deadline stretched to March 31, and then to May 15.

During the spring of 2010, TLC partnered with St. Michael's University Junior School to facilitate a brainstorming session, with students creating an impressive list of ways to help Madrona's cause. As Erika Verlinden wrote in an article in TLC's spring 2010 issue of *Landmark*, the students "left the meeting charged with energy and ready to put their ideas into action through the school's Environmental and Service Clubs." True to Verlinden's article title, the "Youth Help[ed] Plant the Final Seeds of Support for Madrona Farm."[489]

Taking Shape

While engaging in environmental studies at the University of Victoria, Nathalie started a research paper that became her restoration project on the farm. Pointing out that more biodiversity in farms leads to environmental sustainability, she led potential readers to also equate greater biological capacity with truly organic produce. Showing clearly the need for the totality of a natural environment to sustain a healthy farm with healthy food, her paper supported her protest against development in Saanich.

When it was complete, she sent the paper out to land owners in her neighbourhood. One neighbour told her that her notions were completely unrealistic. "You can't talk about sustainability if you don't own the land, right?" he said. "Because sustainability is how things exist over time." She answered his challenge by securing Madrona with a long-term sustainable lease, which will be handed from farmer to farmer within the land trust.

Others contradicted her by saying, "If something isn't arable, you should be able to build on it." She explained that biodiversity buffers are essential to agriculture and that soil can be amended through sustainable farming practices. Pollinators live in these protected habitats. Complex soil micro-organisms thrive in these spaces, as do birds – pollinators of produce as well as consumers of invasive pests. Together, they all provide ecosystem services.

When they received biased and/or uninformed feedback from readers of Nathalie's paper, the Chamberses began to realize the extent of the bias against long-term, sustainable farmland. In creating the Chef Survival Challenge, David and Nathalie built both a fundraiser and an educational tool. They designed the obstacles on the survival course to emulate the obstacles to food security, not only on Vancouver Island but nationally and globally. Monkey bars stand for real-estate prices, hurdles for loss of biodiversity, and a zipline represents soil capacity. This version of the game of "Survivor" includes a lot of metaphors intended to bring together people who have many different ideas about sustainable farmland. Fun and educational, the Chef Survival Challenge is irresistible to the large audiences that gather to watch often-famous people taking part in ridiculously carefree events right on the land where their food is grown. As people leap about in their growing produce, David and Nathalie bond with their community.

When Nathalie and David were in the midst of planning the first Chef Survival Challenge, they received offers of help. Because the chefs wanted to guarantee their source of favourite, local, organic vegetables, they eagerly asked how they could lend a hand. Offering help to Nathalie meant instant admittance to the Chef Survival Challenge, which grew in complexity each year as she added elements to the game. Consistently focused on reconnecting people to the farm, to the land and to their food, the Chef Survival Challenge succeeds wildly every year – both with participants and with the audience.

Nathalie believes that everybody holds a farm in their heart. Everyone has a connection to a farm, a memory sometimes smelly, sometimes sweet, sometimes muddy. Certain sounds, smells and textures take us right back to the place where we felt most at home with nature and its provisional essence. Food both heals us and woos us. "What is the fastest way to someone's heart?" Nathalie asks. "Through their stomach!" When people eat food that comes from Madrona, they receive a simultaneous serving of education. Once they see and feel the land, it changes them, especially when they come to Chef Survival Challenge events. Surrounded by community, they become part of the land, in the spirit of contributing to its biodiversity.

Nathalie appreciates the Chef Survival Challenge for its ability to provide many donors, instead of just one. The event brings the community together. She sees the event as a harbinger of social change, and that social change as an indicator of the level of consciousness that exists at any given time. Although it may seem frustrating to consistently face ignorance, Nathalie points out that it can also be a blessing to be aware of it. If we know it, we can address it.

By dreaming of many donors as opposed to a few, Nathalie hoped to create a revolution. When 3,500 people came together and raised the necessary $2.7-million, they fulfilled her dream. Multiple donors working together on a campaign like Madrona's allows the financial risk to be shared by many. Having 3,500 donors also firmly demonstrates a broad level of support and enthusiasm.

Nathalie used her sense of panic as a motivator when donators began questioning the degree of "perpetuity" possible for Madrona during the TLC strife. At such points, she didn't always feel motivated. At more than one point during the campaign, she said, "I feel as if all my ideas have been dried up – like they've gone swimming in an empty, concrete pool. I know the key is to just keep on moving, even if I have to walk instead of swim."

If she was moving, she was learning.

After the campaign finally succeeded and Madrona's ownership was officially transferred to the trust, Nathalie vowed she'd never run a Chef Survival Challenge again. Ever. However, as the summer came to a close, many of the previous attendees began asking when the challenge would be repeated. David Mincey, former owner of Camille's Fine West Coast Dining, and the auctioneer for each dinner created by the racing chefs, persuaded Nathalie to keep the event alive.

That summer of 2010, Nathalie had just started working as the agricultural program coordinator at TLC. Her life had just started to normalize when, at the last minute, she started to plan the 2010 challenge and her life grew crazy again.

She viewed her process in planning the event almost as an obsession. Although she enjoys only about 40 minutes of it, those 40 minutes always make everything worthwhile. Each year, the moment arrives when people are having such a good time that a certain vibration hums in the air.

Thus, the Chef Survival Challenge worked its way into becoming an integral piece of the overall, post-campaign Madrona model, with the proceeds going into a general fund.

The Bigger Picture

While the Chef Survival Challenge raised $17,000 in 2009, it brought in $7,000 in 2010. People didn't feel as pressed to donate after Madrona had been saved in perpetuity. The decrease in interest goaded Nathalie to further educate. After all, Madrona's success represents just one small step in the right direction.

The vast majority of people seem unaware that we're in a food crisis. As long as the shelves are stocked at Sobey's or Safeway or Tesco, people assume their food source is secure. "The ability to think critically seems rare these days," says Nathalie, "and that's scary because it's a vitally important skill."

The ongoing food crisis nudged Nathalie into creating a non-profit associated with the Chef Survival Challenge, in its fifth year (2012) – the Big Dream Farm Fund[490] (a non-profit).

The Chef Survival Challenge Inc.'s charitable status meant that Nathalie had limited ability to donate. Because corporations can give money to whomever they want, Nathalie created the Big Dream Farm Fund as a corporation. She wanted to give money to the National Trust movement because she believes that reframing ownership is key for the future of the commons. The Big Dream Farm Fund started looking over applications for grants in November 2013 to choose which organizations best fit the mandate, and thus which groups it would fund.

For the first five years, Nathalie did not pay herself anything for her work on the Chef Survival Challenges. The sixth event enabled her to use a humble amount to cover part of her trip to Uganda.

As a result of Chef Survival Challenge, thousands of dollars have been dispersed to farmland protection and sustainable education. One hundred per cent of the proceeds have been given to these causes since the Madrona campaign wrapped up in 2010. Nathalie has given money to *Victorian Naturalist* journal,[491] Food Roots,[492] Share Organics,[493] Lifecycles Project Society,[494] Farmlands Trust,[495] Pender Island Community Farm Project[496] and Horse Lake Farm,[497] among other organizations supporting sustainability and the goal of food security.

Nathalie dreams of planting an annual Chef Survival Challenge in every province and territory in Canada.[498] She envisions the challenges taking root so that they grow a society with 20 per cent of the proceeds going to sustainable farming education, 70 per cent to acquisition via land trusts and 10 per cent to run future challenges. Beyond her dreams of helping Canada become food secure, she's looking farther afield to help the world while also finding help from the world.

While continuing to fundraise, Nathalie is on the hunt for the next bearer of the Saanich Greenbelt Society's torch. Once she's passed it on, she envisions owning 51 per cent of the shares of her corporation (The Big Dream Farm Fund) to ensure that money continues to support only sustainable farmland security. Any person, she says, can take it on. She believes her vision is only a skeleton and invites others to help her flesh it out.

The spirit of rotation is inviting. As soon as people are honoured, they feel capable and part of the community. Recognizing one another as integral parts of a whole helps each of us keep ourselves intact in a healthy, communal body.

In a news release on May 14, after they had saved the farm, Nathalie said, "This project has changed our lives. David and I are true believers that a community is one of the most powerful forces in nature, and if we all stand together, there is no problem too big for us to solve! Our community now has a working model for sustainable agriculture – and food security is on the horizon. Thank you, everyone who had a hand in helping this campaign, and especially The Land Conservancy for stepping up to the plate."[499] Ed Johnson, chair of the Farmlands Trust, responded by congratulating everyone in the community, saying, "The celebration today confirms that protecting the future of our farmland is possible when you hold a dream in your heart!"[500]

PART THREE
Keep It Up

"The point of commons care is to prevent harm before it occurs. And that means learning to 'think like an eco-system.' . . . We come to see natural treasures no longer as merely divisible property but as gifts protected by boundaries we create and honor, knowing that all life depends on their integrity. The shift is underway."

—FRANCES MOORE LAPPÉ[501]

TWELVE : Trust the Commons

Defining the Commons

Portland, Oregon, architect Mark Lakeman took a break from his work and travelled from Central America to Italy. As he sat in the plazas and other community gathering places, he absorbed the relaxed richness of the people – an uncountable wealth missing in many North American cities and towns. He returned to Portland determined to effect change. Creating fun and art in surprising places, at intersections and in squares, he soon infected others with this idea of joy in gathering people together. As the idea of neighbourhoods coming together spread, Lakeman and others formed an informal organization called City Repair,[502] which focused on the idea of village-building convergence, with volunteers transforming all types of public spaces into lively commons. This Portland model is now spreading around the United States, thanks to Lakeman's frequent national talks and his firm, Communitecture, dedicated to designing "beautiful and sustainable places that bring people together in community."[503] Saving farmland is a part of this beautiful sustainability.

When we consider the commons and the movement to save farmland, we must understand the "commons trust." For a quick definition, let's consider its two main opposites: a corporation or a government. A corporation must bow to making profits for its owners and shareholders above all other considerations, and – as we discussed previously – it has no legal personal liability. A government is a political creature with a short horizon, elected by campaign donors and voters, often contaminated by pressures from corporate interests, and with no mandated sense of caring for future generations.

A commons trust, however, is an institution with a legal responsibility to future generations. Its job is to foresee harm and prevent it. Its trustees can "make hard decisions without committing political suicide. They might be appointed by the president . . . but they wouldn't be obedient to him the way cabinet members are. Once appointed, they are legally accountable to future generations."[504]

A website called Remix the Commons takes the definition of "commons" several steps further: "We can identify a commons whenever a community is united by the desire to take care of a collective resource or to create a new one using participatory and democratic norms in the public interest. Many issues can be approached through the commons lens: water, air, forests, oceans and other natural resources; culture, language, computer coding and works of art; the human genome, seeds, and prescription drugs. All these issues and many more can be seen as commons."[505]

The Preservation Benefits of the Commons

In a well-informed, forward-looking world, we would retain our existing corporations and governments, but we would calm their ruffian side with the mandates of a commons trust.[506] Without having to bow to corporate lobbyists the trustees could carry out the commons trust's mandate, which would ensure long-term preservation of nature's irreplaceable resources. (Lobbying pressure is relentless, with a current ratio of 61 corporate lobbyists to one elected representative or senator being common in Washington, DC,[507] and a ratio of 12:1 in Canada.[508]) In the commons model, nature's seeds, for example, would take priority over genetically modified seeds. Access to those seeds for the common people would forever be a high priority. Agricultural land would naturally be included in the mandate of the commons trust, as it is, for example, in Marin County, north of San Francisco, where farmers can choose to sell permanent conservation easements to their choice of two trusts. Farmers receive a payment equal to the difference in land cost between its housing development value and its farming value, and they give up future speculative value. However, they retain the right for future generations to farm it. This permanently protects land use solely for farming.[509]

In 1980 forward-thinking citizens succeeded in creating the Marin Agricultural Land Trust.[510] By 1993 environmentalists succeeded in creating the Pacific Forest Trust following the 1990 Redwood Summer, which saw activists by the thousands chaining themselves to trees in northern California, thwarting bulldozers in a massive protest of unsustainable logging.[511] By 2008 a third of current farmland in Marin County – 40,000 acres (16,187 ha) – had been permanently preserved. As bordering forest land had also been preserved in a trust, this commons solution ensured sustainability through biodiversity.[512]

The Success Rate of the Commons

Commons trusts can play other roles, such as controlling pollution by charging fees that are returned to the common citizen and the ecosystem both in the present and in the future, which prevents the medical costs associated with illnesses that would otherwise arise due to contaminations of water, air and earth. Some may question how any group of people (corporate, governmental or commons) can resist corruption and spoil good intentions. In 2009 political scientist Dr. Elinor Ostrom received the Sveriges Riksbank Prize in Economic Sciences in Memory of Alfred Nobel for demonstrating the positive side of how the commons succeeds while other forms of governance teeter. Ostrom shares the prize with economist Oliver Williamson.

Ostrom didn't start by trying to prove any assumptions of reality. Instead, she collected information in the field and analyzed it. Her book, *Governing the Commons*, proves how users of common property can indeed successfully manage it. Participant users make the rules and enforce them. If we're not involved, and if we neglect to educate ourselves on the issues by reading and talking to a variety of others in the community of our shared interests, we can too easily sit back, put on our blinders and let a small minority steer us off the cliff. Ostrom and Williamson's prize in economics honours the facts proven in their research: "ordinary people are capable of creating rules and institutions that allow for the sustainable and equitable management of shared resources."[513]

The duty of a trustee for a commons involves neither votes nor profits but the preservation of a commons ecosystem for generations to come. In the words of Frances Moore Lappé, sharing an "ecomind" about the commons allows us to "to see natural treasures no longer as merely divisible property but as gifts protected by boundaries we create and honor, knowing that all life depends on their integrity. In this shift, we come to value what we share as much as what we own, what keeps us alive as much as what we exchange."[514]

Do Commons Endure?

How realistic is it to expect a commons to last? Lappé points us to James S. Fishkin's *When the People Speak*,[515] which discusses the farmers in Torbel, Switzerland, who have kept their commons of grazing meadows, forests, irrigation systems and roads functioning well for at least five centuries.[516] As long as a commons is truly shared, and not tipped into a power imbalance so that a small minority dominates or controls it, it should be able to continue according to its mandate in perpetuity.

Can the World Embrace the Commons Concept?

Many believe that change, which is required to help the commons concept grow, takes a long time to come about. But it doesn't. Once a group of people overcome their resistance and experience the benefits of positive change, new ways of thinking spread like fireweed in wind.[517] As shared by the authors of *Powering the Green Economy: The Feed-in Tariff Handbook*,[518] in 2001 Germany began rewarding householders who produced renewable energy. Within a short time, more than 82 other governments adopted their reward system. Frances Moore Lappé notes that in only ten years of rebate incentives, California multiplied its first five hundred solar roof panels one hundred–fold, to cover 50,000 homes. Following this, in the year 2008 alone, double the number grew in ten other states.[519]

Lappé believes that a commons can be accepted by both the right and left sides of the political spectrum. She points readers to Herb Walters, founder of RSVP/Listening Project based in North Carolina, who explained:

> . . . many conservative Christians have held negative perceptions of environmentalism because it's been linked with liberal ideology, secularism, and "environmental extremists." They felt none of this included them.

But a Listening Project in two rural North Carolina counties found that resistance often morphed into action when hour-long interviews with trained listeners offered church leaders a chance to reflect on stewardship of the earth as God's creation and, from there, to reclaim their biblically based sense or responsibility for creation care.[520]

Walters explains, "Activists often see themselves as having the answers. Listening Project changes the rules by treating each person interviewed – even our opponents – as the expert. When we listen to people, they may find common ground or offer solutions. Some then take the next step, working for positive change."[521]

In one of his articles, Walters points out that too often activists polarize, creating camps: "the educated" versus "the ignorant," or the "good guys" versus the "bad guys."[522] Each one of us wants to be heard. If we're told we're the enemy without having a chance to express our concerns in a caring, two-way discussion, we naturally cover our ears and run from the perceived bullies. Some activists base their protest cases on fear-mongering. As the fears preached by these activists ramp up, we run in the other direction. If we feel defensive, we tend to shut down, and the chance for mutually beneficial change flies out the window.

We need to feel welcome crossing into the grey zone between the black-or-white divide. If we can be aware when we're either shutting down or dominating, if we can summon the courage then to either open up or back off, we can enter into productive discussions that help us journey into that vast grey zone of productive possibilities for change. Being open to taking the time to listen to each other equates to allowing time for the seeds of great ideas to blossom into nourishing gardens of future opportunities. If we cannot do this, we cannibalize our hopes for the future – especially our hopes for healthy food forever. Every culture holds an abundance of beauty, compassion, empathy and ethics, which, if embraced with loving attention by all cultures, can grow respectfully together into a beautiful potential for positive change.

Even if we're not feeling polarized, it's easy to feel overwhelmed by the impressive number of possibilities for taking action. When you feel this way, it's important to take a breath, step back, pat yourself on the back for what you do manage to accomplish and take a rest to refocus. Lappé acknowledges this in *EcoMind*'s last chapter, where she lists "thought traps" that tend to cause panic and immobilize us, followed immediately by "thought leaps" that can help us move out of each trap into effective action. Once we refocus, we can gain courage and support by choosing to join a community of other broad-minded people who are brave enough to explore and offer support in that bewildering grey expanse between black and white. Lappé presents us with a summary of positive-thinking, forward-moving organizations, magazines and books.[523] Her list of resources can help us cultivate our interests, honour our needs, nourish our creativity and take meaningful, educated action. We can experience the satisfaction that comes when we surround ourselves with community, and in doing so we can help our planet heal and reclaim the commons.

How a Commons Might Have Saved 370+ Hectares (914+ Acres) of Farmland from the Military

In 2013 the Canadian Department of Defence coerced farmers on over 283 hectares (700 ac) of Class 1 Ontario agricultural land to agree to allow our government to cover their land in concrete and surround it by metres-high walls around a new headquarters and training centre for the highly secretive

Special Operations Forces Command adjacent to Trenton Air Base. As agriculturalist, social activist and author Helen Forsey said, "The growing of food, which sustains life, will be replaced by provisions for war, which destroys life."[524] If this weren't enough, the Department of Defence put extra pressure on Frank and Marjorie Meyers, an elderly couple who had the temerity to refuse to give over their 89 hectares (220 ac) of prime, Class 1 farmland. It had been in their family for over two centuries. Somehow, the couple was pressured to sign papers. As of April 2014, the Department of Defence was the registered owner – even though it hadn't paid anything to the Meyerses for the land that had been theirs for multiple generations.[525]

When a government refuses to explain its motives to its own citizens, how do we know its intentions? Does Canada's top-secret "security" intend to protect all its citizens, or only its wealthiest corporate supporters? Will the land taken be used to protect resource businesses? Or for counterinsurgency against civilians who dare to complain? In October 2013, Conservative MP Rick Norlock told *Maclean's*, "I guess the good of the many, in this particular case, outweighs the good of the few."[526] Has the food from all of that prime farmland been providing "good" for only a "few"? When only 5 per cent of Canada's land is arable (never mind prime, Class 1 soil) why couldn't the military have chosen to locate itself on any of the other 95 per cent?

Meanwhile, the Meyerses are still living in their house across the street from the farmland. In the spring of 2013, Frank was allowed to plant one last crop of corn to sell as cattle fodder, although he'd been forced, under threat, to sell his own cows. The autumn was too wet for the corn to dry, so Frank decided to wait until May of 2014 to harvest it. In early January 2014, the government brought in heavy equipment to demolish Frank and Marjorie's barns. Social-media commons leaped on the case, and within an hour people who had never before been activists arrived with protest signs, a trailer, firewood and food to mount 24/7 pressure. As Forsey so eloquently points out, this aroused group of common people suddenly understood that this "physical and spiritual ravaging" is akin to "a modern David with a social media slingshot, challenging a military and political Goliath."[527]

The Meyers Farm Facebook page has become a commons in itself, where the latest updates continue to be available. On August 14, 2014, the Save Frank and Marjorie Meyers Farm Facebook page explained what happened when Frank tried to follow through on his corn harvest:

> Months later after a brutally cold and snowy winter and late spring Frank prepared to harvest his crop in May as soon as it was dry enough. On the 21st of May the DND crept up the back side of the field in the early morning with trucks and excavators to place a barricade of rubble torn up from Frank's barnyard at the entrance of the field. A week later on May 28, while Frank was away from the farm, 84 year old Marjorie Meyers heard a commotion and turned to see the destruction begin. She stood alone, in tears, helplessly watching the DND tear down the Meyers' barns.[528]

The Facebook page continues to encourage action, for as its contributors say, "We all have a voice and now is the time to use it and INSIST that the government do their job properly and relocate this expansion instead of expropriating prime farmland. We have already lost over 8,000 Canadian farms since this present government took power."[529] We need to start taking back the farmlands of Canada and start growing more Canadian food instead of importing food that can be grown here. The Crown owns over 85 per cent of the land in this country, and the DND in particular owns thousands of hectares of mothballed bases sitting idle.[530] However, to have prevented this tragedy, more people needed more awareness and more powerful leverage. If this farmland had been part of a commons trust, it could not have been expropriated.

As farmland converts to coffins and the rate of hunger mounts, a flag – obviously not erected by the Canadian military – is waving more and more vigorously, signalling us to take Elinor Ostrom's research to heart. Ostrom emphasized the importance of both accountability and enforceability in building trust to grow an enduring commons. The prevailing ethic of the commons contributes to the common good.

Current Growth of the Commons

Order of Canada recipient, professor and author Heather Menzies points out in her *CCPA Monitor* article, "Commons of the Past Inspiring Model for Today's Progressives,"[531] that the word "common" had its origins as a verb. People "commoned," coming together to share in working bees, to maintain roads, farms or fences, build homes in a community or set limits on how many sheep, goats or cows a family would be allowed to forage in common hilltop pastures so as to avoid overgrazing. In the preface to her book, *Reclaiming the Commons for the Common Good*, Menzies further explains, "The word common originally meant 'together-as-one,' 'shared alike' and 'bound together by obligation.' This togetherness was not only with each other, but with the land itself. It's time to reclaim the commons, first as memory and heritage and then as practices and the capacity to once again live together-as-one with the Earth."[532] Menzies explains how this is now happening.

Inspired by Elinor Ostrom's goal to bring the commons into the modern era, economist Shelagh Huston and permaculturist Heide Brown formed part of the board of the Amazing Grace Ecological Society (AGES) on Gabriola Island, near Nanaimo, BC. In 2005 AGES bought a 10.5-hectare (26-ac) former goat farm, and the group met twice a month to navigate land-use bylaws, governance models and fundraising strategies for expenses. They submitted an application to the Islands Trust for a new zoning bylaw "where sustainability, community and agriculture meet."

The result? "In 2010, two new Islands Trust bylaws were passed. One, #258, officially established and defined for the first time in Canada a 'community commons;' the second, bylaw #259, officially recognized the Gabriola Commons as a Comprehensive Development Zone in its own right. In January 2011, these bylaws were approved by the provincial government."[533]

The Gabriola Commons now has a commons coordinating council, which holds monthly meetings open to everyone and supports teams that manage the farm, plan special events, and maintain the grounds, trails and green spaces. In addition, a covenant team welcomes all input to combine both nature and culture in the commons. All work together to hold the commons accountable to future generations. The group's guiding principles are based on the fact that "The Gabriola Commons is an ongoing, reciprocal relationship; the people care for the Land and the Land nurtures the people."[534]

Local and quick decision-making ensures that a commons like the Gabriola Commons is successful in its actions – as opposed to governments or corporations, which are often located great distances from the places affected by their decisions and whose decisions often take years to come into effect. As the local, commons community is involved daily in its own environment, it has the greatest understanding and the greatest impetus to take care of it and to share it. As we connect with Earth, we increasingly desire to relate with it, rather than to gain (political or corporate) control over it. Our growing natural connection with our world can create the beautiful, natural greening effect of being part of a commons.

Examples of Modern Commons

The desire to share any sort of commons naturally builds a momentum of caring and preservation. Examples of modern commons include global information commons such as the Occupy movement,[535] cultural/heritage/environmental commons such as the Idle No More movement,[536] many of the co-operatives worldwide, or Lakeman's beautiful Communitecture.[537] This model of generous sharing of information for mutual, common benefit has been growing. A variety of universities – including MIT,[538] Yale,[539] the University of Washington[540] and Carnegie Mellon University[541] – offer many of their courses and lectures free online for people who wish to learn but who aren't concerned about receiving degrees. Several of these are listed as Massive Open Online Courses (MOOCs).[542] Commons sharing also occurs at the individual level, as demonstrated by Salt Spring Builder, a Salt Spring Islander and skilled carpenter who has shared over 30 years of expertise in a compilation of videos that provide tips on house repairs and renovations.[543]

For the same reason we exercise to keep our physical bodies strong, healthy people worldwide have practised a commons-style manner of counselling to regularly and safely release emotions in order to think more clearly. Called co-counselling, this more than 50-year-old support system allows clients and counsellors to spend equal time in each role. People work in pairs or sometimes small groups, with the client discharging feelings through laughter, tears, yawning or non-repetitive talking in the loving, respectful, confidential presence of the counsellor. The premise assumes that each one of us is the best person to solve our own problems as long as we are able to express the feelings that we've saved up from a time when we did not feel safe to express them. In the safety of a confidential commons, sharing expression time equally, we are encouraged to release, to re-evaluate and to then move forward.[544]

People who wish to grow in agroecological knowledge by practising organic farming while travelling around the world can explore World Wide Opportunities on Organic Farms. An international collection of WWOOF groups co-operates in sharing common responsibilities for managing a combined portal website to ease and encourage participation.[545]

A commons approach involving a community of clothing workers in Bangladesh would never force its employees (being themselves) to work in unsafe conditions. Nor would a commons local to the Amazon allow the burning of nearby rainforests in order to export cheap produce far away from its own hungry community. A corporate boss, located a great distance from the area affected by her decisions, tends to be less connected to that place and therefore less caring than she would be if she were "on the ground." If that same person were instead an active, questioning member of a commons, she would be part of a group gaining knowledge about the community in order to efficiently make the right decisions. The learning in a commons spreads quickly and in a caring, supportive manner, serving both current and future benefit. The Occupy movement's actions, for example, reflect an international citizenry starving for alternatives to bureaucracy and ready to leap into involvement at the community commons level. Such a community, with its inside knowledge, understands best its resources, can most easily discuss the limits of its local environment and establish rational rules, is available to monitor the effects of decisions, can supervise and guide those who break rules, can work together in conflict resolution – and will seek outside authority support only when the above has already happened and didn't work.

As Heather Menzies notes, outside the paternalistic-seeming

conservative and the maternalistic-seeming liberal political choices, "the commons model offers a hopeful third choice: re-enfranchising people as responsible co-participants in the governance of the larger habitats that sustain them, including in their individual lives. In this approach, we are not left to help ourselves on our own, nor are we waiting to be rescued and looked after – but come together in various communing and common cause arenas to share looking after the larger whole of community and habitat."[546]

The Madrona Farm model, with the land trust owning the land and offering long-term sustainable leases to farmers, is also a commons that guides farming practices and welcomes both paid and volunteer workers. This commons, having an intimate connection with the soil, the pollinators, the neighbouring trees and plants, mandates organic, biodynamic and permaculture practices – illustrating agroecology at its best. Sharing their knowledge in commons fashion, the Chamberses hold community events like the Chef Survival Challenge and Mushroom Volunteer Day, during which volunteers help inoculate two hundred logs for shiitake and oyster mushrooms, in the process learning how to propagate mushrooms in their own backyards or farms.[547]

Over the past half-century, we have become numbed to the point that we have allowed ourselves and others to commit acts of chemical warfare against nature. Our treatment of nature demonstrates the extent to which we have separated ourselves from our communities and our base of healthy food. The growing popularity and success of the commons provides an opportunity for each of us to reconnect – and to heal ourselves physically, mentally, emotionally and socially.

"Food brings people together on many
levels. It's nourishment of the soul and body;
it's truly love." —GIADA DE LAURENTIIS[548]

THIRTEEN : Share Mutual Success

Each family facing the stories of farm loss in chapter 1 of this
book saw beyond the tragedy of the moment. They propelled
themselves into forward-moving visions, with ethics insisting
that they care both for the current community and for future
generations. They embraced other possible solutions, remaining flexible as other paths – or roadblocks – opened and closed
along the way.

The changing nature of politics and our own learning keeps
offering us opportunities to discover more about the wisdom
inherent in nature. If we keep our eyes and ears open – and our
hearts full – we will grow to understand how ecosystems sustain life in all ways. Hearing one another's stories will nourish
our ability to preserve healthy farmland and its bio-network
forever. Sharing mutual success helps nourish us all.

Arran Stephens: Paradise Regained

After leaving Goldstream Berry Paradise as a child, Arran
hugged the memory of it in his heart, inspiring his wife Ratana,

their children and grandchildren to create a new paradigm for both business and farming. As they saw the price of farm-land climb out of reach for young farmers, they purchased two, organic grain and legume farms. These they offered to skilled farm families to cultivate sustainably on a co-operative basis, sharing the crops with them on a two-thirds/one-third, win-win basis. After buying several thousand hectares in Saskatchewan, they watched with joy as two of their family's sons – whose hearts were in farming but who were stuck in the oil fields – each flew back to fulfil their goals to farm and raise a new generation of farmers. The family's organic enterprise now employs a staff of hundreds, supporting farmers by the hundred across Canada, the US and South America.

When asked how he manages the challenges of Nature's Path, a packaged cereal company dwarfed by multinationals that are thousands of times bigger, he says, "Firstly, by not selling out. Secondly, by structuring our succession to involve the next gen-eration on a merit basis, and ensuring that our (my wife and I) ideals and commitment are fully engrafted into the DNA of the company, so that it will carry on long after we have become organic compost! We are nimble players; we dance under the legs of elephants. But although Nature's Path may be minuscule by comparison, our rate of growth is enviable, and we love what we do."[549]

Continuing to encourage new farmers to farm organically, the Stephens family is meeting the rising demand for pure food. They have not sold out their highly successful organic cereal business to Wall Street or to junk-food conglomerates. Time and time again, they have refused. "One day several years ago," Arran said, "I received a phone call from Kellogg at 10:00 am, and another phone call from Kraft two hours later. The dollars offered were staggering."

"We have turned down hundreds of millions of dollars in private-label business," Arran said of the supermarket brands that have been pestering him to sell their cereal under their label, "because we want to protect the integrity of our brand. When you get into generics, the absence of a brand means price becomes your brand, and it's a downward spiral. We want a legacy brand."[550] He continues:

> We asked ourselves if all the money that was offered was going to improve the character of our children, and would it improve the lives of our many employees and stakeholders, the communities and economies that we worked in, the organic movement that we have served and helped build over so many years. And we decided no.
>
> We would run our company better independently, and make it strong enough to survive and thrive into the next generations. The offers still come every week, but they find themselves in the round file, and the voice messages are deleted.[551]

Meanwhile, Arran and Ratana intensely cultivate 0.2 hectares (0.5 ac) of organic vegetables alongside one hectare (2.5 ac) of berries and fruit trees. Their son, Arjan, manages a large community garden in Richmond, BC. They all donate their produce to the SOS Meditation and Ecology Centre,[552] which provides a free meal for 80 to 90 people every Sunday. Besides building other Richmond gardens, they provide three annual grants to keep community gardens established in areas where they are needed most.

Although Arran realizes that he can neither turn back the clock nor restore his childhood farms to their original beauty, he has stayed true to his father Rupert's original vision of sustaining the soil, bettering it wherever possible, permeating his children with it, keeping his business proudly independent and family-owned, yet sharing the proceeds generously. He hopes "that it can be said, when this bod is no more, the soil was left more fertile than today."

Rupert later wished he had been able to pass on his two family farms, but he shared lessons learned from that loss in one of his hundreds of songs. His son and grandchildren learned from those lessons, as a result.

> This earth is mine to treat with love, right here beneath my feet;
> I know that on the summer dust, the steam of rain is sweet;
> And when I'm weary, old and worn, I know that I can say,
> "I'll pass it on to other hands, more fertile than today."
> It waits for me with loving arms, this Earth is mine that day.[553]

Harold Steves: Anger Igniting a Half-century of Positive Political Action

After the Richmond City Council's secret "zoning had gone through, one by one, the farms disappeared."[554] Each year, for more than a decade – despite protests – developers continued flipping six thousand hectares (14,826 ac) of BC farmland[555] into industrial parks, subdivisions, golf ranges, super-ports, highways and pipelines.

Mike Guichon, representing the Delta Farmers' Institute in 1973, pointed out that the soil quality deposited by the Fraser River in this broad estuary is "some of the finest on the continent." Slathering it with concrete in the prior years resulted in memorable "tragedies committed in the Delta farming community."[556]

If the flames of anger can be identified and fanned thoughtfully, they can fuel powerful, effective action. Originally a farmer, scientist and teacher, Harold understands the political

268

nature of farming. First serving as an alderman from 1968 to 1973, he next won the vote as a New Democratic Party MLA in the BC Legislature, serving from 1973 to 1975. During his time as an MLA, he co-wrote the BC Land Commission Act which came into effect on April 18, 1973. Between 1974 and 1976, regional districts and member municipalities held public hearings to gain enough input so that BC's Agricultural Land Reserve could begin protecting 4.7 million hectares (11,613,952.9 ac), 5 per cent, of the province's arable land.[557]

Delta's current mayor, Lois Jackson, a fresh alderman when the ALR was born, points out that despite the stark polarization of left (NDP) and right (Social Credit) politics of the day, the ALR managed to endure both sides of the political spectrum for over 40 years because its captains believed so strongly in farmland protection. "In retrospect," she says, "had the Agricultural Land Reserve not been in place, Delta as we know it would look like Richmond. This flat land we have here in Ladner and Tsawwassen is wonderful to build on – but also contains the very best farming soil in the province. The one thing we have to remember is that only 5 per cent of this very huge province can sustain any type of crop. Without the ALR, we would have far less in production than today."[558]

The ALR did indeed erect stop signs. At the very least, developers had to take notice of the food base before trying (often unsuccessfully) to head down the road of exemptions. As mentioned in chapter 7, the provincial government in office at the time of this writing – without any consultation – dropped its sledgehammer on 89 per cent of the farmland protected by the ALR so that the doors could be opened for northern gas and oil development.

Knowing that he would have to continue fighting, and understanding how food intertwines with all aspects of the social and physical environment, Harold has held a variety of preservation office positions, defending farmland, heritage and the environment his entire life. While continuing to farm in every spare moment, he has served (and in many cases, still serves) Richmond on Metro Vancouver's board of directors, as chair of the Parks, Recreation and Cultural Services Committee. He is also a member of the Finance, General Purposes, Planning and the Public Works and Transportation Committees. In addition, he serves as council representative to the Agricultural Advisory Committee, the Olympic Business Advisory Committee, the Richmond Farmers' Institute, the Sea Island Community Association, the Steveston Harbour Authority Board, the Britannia Heritage Shipyard Society, the Richmond Athletic Commission and the 2009 Richmond BC Seniors Games Board.[559]

Harold Steves and his family are dedicated to keeping us all well-informed. At their Steveston Stock and Seed Farm website,[560] engineer/farmer son Rob connects people to the Nova Scotia Pinnacle Farms' "Farm Facts or Fiction" animated site.[561] The Steveses also include links to many other websites and stories whose titles alone entice readers to learn more. These are a few samples:

- Eat Wild: Getting Wild Nutrition from Modern Food;[562]
- Seeds of Diversity;[563]
- "Small Farmers: Vital Work, Slim Wages";[564] and
- "Why Urban Farming Is the Future".[565]

The Steveses share captivating book titles on their site, too, including *The 100-Mile Diet: A Year of Local Eating*'s positive and forward-moving sequel by co-author J.B. McKinnon: *The Once and Future World: Nature as It Was, as It Is, as It Could Be.*[566]

Beyond the Steveses' website, Harold has given many inspiring talks, drawing on his education as a geneticist and his experience as a farmer. One of several examples of his presentations available on YouTube is "We Want to Know What's GMO."[567] In this speech, given on October 12, 2013, he clearly explains the dangers of genetically modified seeds during a Vancouver march against Monsanto.

In the summer of 2013, Harold Steves reflected on his almost half-century of political activism. "I don't think it would have happened without me getting angry when my dad was turned down for his barn," he said.[568] While growing vegetables from seeds preserved for more than one hundred years, while broadly marketing those heritage seeds, while raising grass-fed beef for customers who must reserve it six months in advance, and while co-parenting five, community-minded children who also care about farming, Harold has been serving the Richmond City Council so long that he's been given two, long-service awards.[569]

During an interview with *The Globe and Mail* in the autumn of 2012, Harold said, "I keep pinching myself every now and then. Did I do that? I must have been crazy." Justifiably, he laughed "a very satisfied laugh."[570]

Madrona: From Harsh Reality to Nourishing Success

Nathalie admits that David's uncle's harsh words, although tough to hear, became the propulsion that lifted the Save Madrona Farm campaign off the ground. "We had to find a way to make all that possible," David said. When Grannie died in 2002, David's family let him continue farming on the land because of his dedication to wholesome food, good cooking and farm restoration. He knew he had no tenure, and he knew that he could be kicked out at any moment.

Madrona was soon producing thousands of kilograms of

food annually, selling it directly from the stand to a list of over four thousand enthusiastic customers, including individual patrons; Food Roots (the local co-op and distributor for organics);[571] Share Organics (a bulk-buying group and home-delivery service);[572] Island Chef Collaborative (a community of food professionals acting to ensure regional food security, preserve farmland and develop local food systems);[573] and numerous restaurants, oyster bars, chefs and golf-club banquet halls.[574]

David and Nathalie's farm had become a sustainable, nature-friendly ecosystem, an example of restored farmland that city people could see and taste. As urbanization closed in on the containment boundary that surrounds Blenkinsop Road and Mount Douglas Park, Madrona's farm property remained the lone stalwart as the sole undeveloped slope on the southwest side of the mountain. Many experienced Madrona Farm as the hub for their nourishment. Thousands of eager customers had learned simply by commuting through the valley daily how farmland in the middle of a city neighbourhood is far more valuable as a community-protected asset than as a tax-generating subdivision. The root of a community's health is in the food that nourishes it.

Over and over again, customers asked if Madrona's produce was all organic. "Yes, but we don't farm this way to fit into a niche market," David and Nathalie would say. "We grow our food organically because chemicals are not a food group and are completely unnecessary for growing high-quality food. Understanding and observing natural rhythms of insects, birds and weather are far more valuable to a farm than anything a bottled chemical solution offers."[575]

By this time, neither David nor Nathalie could see themselves walking away from their love of the land and their associated community. Nor could they see themselves abandoning the idea of building consumer awareness of the necessity for saving farmland. They knew that if any of the land ended up on the open market, Madrona as a food model – and its agricultural and ecological value – would be lost forever. Each new homeowner on Madrona's land would have a lovely estate with mountain or ocean views, depending on which parcel they bought. But the death of the farm would mean another funeral for healthy food, another casualty decreasing Vancouver Island's food security.

"If you walk into a school and ask any graduating student to buy a piece of farmland in the middle of the city for $100,000 an acre, then buy $200,000 in equipment and start growing food at their own risk for the community, they will laugh at you," said David.

"But protect that same farmland," said Nathalie, "and you have now created opportunity. It's easy to inspire farmers if the capital cost of the land is removed. The goal here is to produce more food. And it is our job to protect what is left for future generations.

"I looked at co ops. I looked at shares. I looked at all these different ways of farmland access to preserve ownership. Then I looked into The Land Conservancy."

At that time, TLC was protecting farmland for farming, although that aspect was later removed from its mandate. (Bill Turner's new National Trust for Land and Culture [for Canada and British Columbia] picked up where TLC left off and began treasuring farmland in 2013.[576]) People who like to eat good food understand that farming encompasses everything – the land with its accompanying soil, culture, heritage, food, people, animals, insects, birds, plants and trees. They understand that all are interrelated and dependent on one another to maximize mutual benefit.

People now question the long-term durability of both provincial and municipal "protections" for farmland. They have seen political corruption surface too often, resulting in agricultural lands being "released" from covenants. Corporations, as well as selfishly powerful individuals, can smoothly manipulate viewpoints so that their short-term interests can overrule the long-term needs of future generations. Those of us who wish to build a trustworthy bank of positive solutions for saving farmland know that each locality has different challenges with different talents. We, therefore, understand the importance of keeping our minds open, of continuing to learn and of sharing results. Awareness of the sensibility of the commons model (historical and newly evolved) has gradually begun to resurface.

As we remain open to sharing, we also remain open to supporting those who are knowledgeable, talented and willing to take on leadership roles from an enduringly ethical base. Madrona's slopes sprout year-round with vegetables and young fruit trees, supported by 32 different types of birds and many species of bees, while the frogs work hard in Madrona's pond. Madrona Farm tours simultaneously engage and educate the community, especially young farmers. Madrona demonstrates that a farm is an ecosystem embracing nature, culture and heritage. The goal is to embrace and encourage as much of the natural ecosystem as possible while providing good food to the community.

"The greater danger for most of us lies not in setting our aim too high and falling short but in setting our aim too low, and achieving our mark." —MICHELANGELO[577]

FOURTEEN : Keep Surging Forward

Focus on Heroes and the Critical Balance

Recalling the "people paradox" (that people are both the problem and the solution), we know that as we surge forward in saving farmland, we will face people trying to tug us down into a spiral of seeming impossibilities. It's crucial to remember that we have plenty of heroes in our families, neighbourhoods, communities, provinces, states, countries and the world. We can make the choice to focus on them in order to gain strength. Just by being themselves and modelling successful possibilities, they offer us their helping hands.

As we move forward toward goals of biodiversity and food security, we find ourselves continually walking a critical balance between the powers that grab good seeds, good food and good land from us, and the powers we wish corporate execs would genuinely demonstrate: love and compassion for good seeds, good food and good land.

To generate lasting change, we must embrace the kind of love that reaches for every biodiverse connection and the type of power that energizes every respected individual. Reaching for true heroes, whether in human form or otherwise, and listening deeply to their stories can provide that most helpful kind of loving power.

Bill Turner

While travelling around BC with Bill Turner,[578] Nathalie learned much more than just the status of sustainable agriculture in our province. Observing his constant focus on high goals even as he responded to personal attacks, she learned a great deal about how to stand up in seemingly impossible situations. As she watched Bill persist through adversity, she understood much more thoroughly why one of Bill's heroes is Michelangelo. From Bill, Nathalie learned that if we set low goals, we can be pretty sure to reach them, but if we take greater risks by setting much higher goals, we'll work much harder and find out how much farther we can stretch.

Bill's many years of experience as a constable and detective with the Saanich Police Department taught him how to transcend problems, overcome obstacles and see beyond limitations. A visionary who is full of integrity, he serves as a member of the executive council of INTO, a member of the board of the Pacific Coast Joint Venture, the Canadian Inter-Mountain Joint Venture and the American Friends of Canadian Land Trusts. Persistent and quietly replete with integrity, he is a past board member and past chair of the Land Trust Alliance of BC, and a past board member of the Saanich Inlet Protection Society and the Provincial Capital Commission Greenways Committee.[579] With a rare level of altruism, he is always thinking of the common interest, balancing nature's and human needs.

Bill founded The Land Conservancy in 1997. While running it, he took a lot of risks to save parkland – and to save farmland, too. He was attacked, but he rose above the storms. In the 15 years he directed it, TLC was able to protect more than three hundred properties. To do this, he grew the TLC membership to over eight thousand, and he inspired tens of thousands more people to volunteer and donate. Although he retired from TLC, he continued to dedicate himself to conservation on an even larger scale by founding The National Trust for Land and Culture,[580] with its mandate to "be a collective of 14 independent, not-for-profit, charitable organizations, that works in all parts of Canada to engage people in protecting and caring for sites of ecological, historic, cultural, agricultural, scientific, scenic and compatible outdoor recreational value, and that incorporates public education in all of its work and activities. Member organizations of the collective will be membership-based and governed, and will work in a professional, businesslike and non-confrontational manner to achieve their objectives."[581] Members are also aligned with the International National Trust Organisation and follow INTO's principles.

Bill has been awarded the Order of Canada, as well as an Honorary Doctorate of Laws from the University of Victoria. In 2012 he was also presented with the Queen Elizabeth II Diamond Jubilee Medal, awarded to people for positive contributions to a Canadian province or community.[582] He clearly demonstrates his motto of stretching for much more distant goals than just the accessible ones.

Nancy Turner, Mary Thomas and Sellemah

Nathalie's face glows whenever she mentions her favourite professor. "I'm absolutely sure," she says, "that a day in heaven begins with botanizing with Nancy Turner. Personally, she is someone that I have striven to be like." Hearing why Nancy is one of her heroes includes stories of two other amazing women.

Nancy Turner is a professor of environmental studies at the University of Victoria, specializing in the inter-relationship between people and plants (ethnobotany) and researching how people know about and understand their ecosystems (ethno-ecology.)

She has spent more than 40 years collaborating with First Nations Elders and cultural specialists to document, save and promote traditional knowledge of plants, Indigenous foods, materials, medicines and their habitats in northwestern North America.

By learning the roles of plants and animals in Native stories and ceremonies with their specific words and beliefs, she has helped save Native languages and vocabulary. Dr. Turner points out that the knowledge of Indigenous specialists has often been overlooked in the past, yet it's critical for us to learn from their thousands of years of observation and sustainable interactions, if we, too, wish to regain a sense of long-lasting, caring management.[583]

The titles of many of over 20 books that she has written, co-authored and/or edited (as well as the papers she has worked on) reveal her passion, as they are such clear statements in themselves. For example: "'We Might Go Back to This': Drawing on the Past to Meet the Future in Northwestern North American Indigenous Communities."[584] In the article "Sustained by First Nations," Nancy and her co-author Patrick von Aderkas share recipes and uses of local, "wild" food, in the process revealing the fact that European newcomers learned from – and were often saved from starvation by – the Indigenous peoples.[585] She effectively channelled her frustration with external forces that block Native communities from sustainable management when she co-authored "Blundering Intruders: Multi-Scale Impacts on Indigenous Food Systems,"[586] so that the common reader can take a first step into education and understanding. She also encourages meaningful next steps.

In *"To Feed All the People": Lucille Clifton's Fall Feasts for the Gitga'at Community of Hartley Bay, British Columbia*, Nancy and her co-authors discuss the Thanksgiving feasts offered by matriarch Lucille from the 1920s to the 1950s in Hartley Bay (mid-northern coastal BC,) where "they served an array of traditional foods, including cambium of hemlock (*Tsuga heterophylla*) and amabilis fir (*Abies amabilis*), edible seaweed (*Porphyra abbottiae*), Pacific crabapples (*Malus fusca*) and highbush cranberries (*Viburnum edule*) in whipped eulachon grease, many different fish and shellfish dishes, and a variety of other dishes from the marine and terrestrial environments of Gitga'at territory. Today, as traditional food is increasingly recognized as vital for Indigenous peoples' health and well-being, Lucille's teachings are as important as ever, helping her descendants to maintain their resilience, self-sufficiency and cultural identity in the face of immense global change."[587]

Native peoples – and Nancy Turner – understand the vital use of storytelling in preserving and encouraging sustainable and healthy use of Earth. However, she makes clear that she believes in offering readers a road out of the dilemmas we have created when she and the other panelists and associates on the International Boreal Conservation Science Panel wrote "Conserving the World's Last Great Forest Is Possible: Here's How,"[588] and when she co-authored the paper, "Ecosystem Services and Beyond: Using Multiple Metaphors to Understand Human-Environment Relationships."[589]

Nancy's most recent – and mammoth – collection of work amounts to a thousand pages in two volumes, sharing Native wisdom about plants accumulated over millennia from southern Alaska through BC to northern Oregon. *Ancient Pathways,*

Ancestral Knowledge: Ethnobotany and Ecological Wisdom of Indigenous Peoples of Northwestern North America.[590]

In 2004 when she was honoured with the title Distinguished Professor at the University of Victoria, Nancy said, "It's a total surprise to be rewarded like this for doing what I love to do. It really makes me appreciate being a part of this university."[591] Among many other awards she has been granted, by July 2009, she was awarded Canada's highest civilian honour for lifetime achievement – the Order of Canada.[592]

Because she so highly values working in the field, it's natural that Nancy would plunge her students into fieldwork as soon as possible. As Nathalie explains, during her studies with Dr. Turner, "The ethnobotany trip into the Shuswap First Nations was absolutely life changing. Three vans took a number of us students into the Okanagan (about five hours' drive northeast of Vancouver) where we met one of the legendary First Nations women that Nancy has talked about for years in all my classes, Dr. Mary Thomas."

Born in 1918 in Salmon Arm, BC, Mary Thomas spent many of her younger years with her grandparents, speaking the Secwepemc language, and naturally understood her culture, "based on a complex and interdependent system. Practical needs, spirituality, social/political organization, kinship, and nature were all interrelated and interconnected."[593] Ripped from her rich heritage and shipped off to residential school in Kamloops, she was suddenly forbidden to speak her language and was taught that her spirituality was taboo. After an early life of marriages to ill or abusive husbands with whom she produced 15 children, she moved to Kelowna with a more supportive, life-long partner. While helping prepare traditional materials for her youngest daughter's school classroom, she finally rediscovered her life's passion and began her "journey to healing."[594]

From 1970 onward, Mary began interviewing Elders, learning more about her culture, making baskets, moccasins and mats, tanning deer hides, and, within seven years she had returned to the Shuswap where she helped create the Salmon River Roundtable, dedicated to restoring the environment. By following her true passions, she eventually worked on forestry committees, advised her community, instructed at Okanagan College and assisted Dr. Nancy Turner with her traditional plant research.

> Whenever Mary spoke, she conveyed feelings of spirituality and sincere love and everyone who listened could sense her closeness to nature, her years of wisdom and the connection she had to her heritage. . . . The [Switzmalph Cultural] Society currently operates a heritage village near the Salmon River mouth where events are held to celebrate traditional culture. . . . In 1998, Mary was one of the keynote speakers at an ecological restoration conference in Victoria that helped energize this movement. To a rapt audience, Mary spoke with simplicity and humility about the devastation she had witnessed and the intense need for everyone to work together to heal the land. . . . Thankfully, the series of lectures on native culture that Mary gave at the University of Victoria were published as *The Wisdom of Mary Thomas*.[595]

She has received many awards.[596]

In 2001 Mary told Nancy Turner that "as a child she used to go down to the mouth of the Salmon River where it flows into Shuswap Lake to dig wapato and water parsnips from along the banks of the river. The water parsnips were eaten raw, while the wapato were placed into baskets to be cooked up later. Now, there's not one plant left down there. Let alone a cattail where the birds used to sing beautiful music. You don't hear that anymore."[597]

Between dams, dikes, irrigation ditches, urban sprawl and introduced agriculture, the wapato had all but disappeared.[598]

One of Nancy's grad students started cuttings of one of the few that could be found – revitalizing it during her thesis program. "This was typical of the sorts of stories floating around all Nancy's classes," Nathalie said. "I found it not only amazing, but in fact life changing to experience them. Now, within the youth of the Shuswap, there is a real movement to collect that knowledge."

Throughout her lifetime, Mary Thomas fought hard to protect the rivers for the salmon run, founding the Salmon River Watershed Restoration Project. She, like many other Secwepemc people, felt a strong obligation to care for the land. "Man is supposed to be the protector," she told Turner. "Humans were given the responsibility to protect the goodies that they created on this mother earth and I'm afraid that we're not obeying that. We're losing."[599]

As Nathalie talked about Mary Thomas and Nancy Turner, her eyes sparkled. "If you took any of Nancy's classes, you would hear the respect and deep appreciation she held for Dr. Mary Thomas's knowledge and how it was now being taught to the youth."

A tragedy marked the beginning of Nathalie's class's visit to Dr. Thomas and the Shuswap: Nancy received a call on their way to the ferry. Mary had just lost two, close, younger relatives in a car crash, and when their father heard the news, he had collapsed after suffering a non-fatal heart attack. Of course, Nancy prepared to cancel the trip. But Mary Thomas insisted that they come. Well into her 80s, she felt compelled to pass her knowledge on to the students while she still could.

"To my surprise," said Nathalie, "the mighty Dr. Mary Thomas was a tiny lady. Wearing tiny shoes, she took us up Fly Hills [an area that has two hundred kilometres (125 mi) of trails, and includes forest, subalpine meadows, small mountain lakes and scenic lookouts] just west of Salmon Arm. As we climbed up the mountain, I could hardly keep up with her. She took us to a lake famous for its loons. As we sat quietly together, Mary said, with tears in her eyes, 'So many of the plants I collected with my grandma have disappeared. Right now, so have the loons.' The silence and sense of sadness seemed as high as the peaks around us.

"All of a sudden, there was a wailing in the skies. A massive flock of loons struck the lake. I felt lit with electricity in the most profound learning experience of my life – not in my head, but in my heart. These birds performed – in the spiritual way of the First Nations tradition. Nancy and I sat side by side, hand in hand, in a funeral that seemed to just keep on going for seeming days, although we didn't stay that 'long' (in European measurement of time) at all."

As Nancy shared her love of First Nations culture and plants with her students, Nathalie met others, including a very special Coast Salish Elder, Joan Morris – Sellemah, as she prefers to be called – of the Songhees Nation (from the Chatham Islands, opposite Oak Bay in Victoria). As a child, Sellemah was thrust through experimental surgeries at the Nanaimo Indian Hospital and was interred in the Kuper Island Indian Residential School. But she held on to the precious time she spent with her grandparents and great-grandmother, growing a deep understanding of the plants and animals of her beloved native islands. She took her experiences and generous nature and became a caregiver at hospitals and later an advisor in multiple cultural, environmental, health, social justice and truth-and-reconciliation initiatives. Sellemah models caring in the deepest sense of the word. In a celebration of traditional foods among Vancouver Island First Nations, Sellemah was listed as a "champion."

Reflecting on her experiences, meeting Mary Thomas and Sellemah, Nathalie said, "When Nancy taught me so personally about First Nations' respectful treatment of nature, she handed me the shared torch that has led to my life of conservation."

Elizabeth May

"One of the world's most influential women," according to *Newsweek*, Elizabeth May has been an environmental activist for over four decades. Although the following list barely touches on the many awards given to her, she has been honoured with the Outstanding Achievement Award from the Sierra Club, the International Conservation Award from the Friends of Nature, the United Nations Global 500 Award, the award for Outstanding Leadership in Environmental Education by the Ontario Society for Environmental Education, the Harkin Award from Canadian Parks and Wilderness Society and the Couchiching Award for Public Policy Leadership. She was also named one of the world's leading women environmentalists by the UN.[600] In 2011 the Saanich and Gulf Island riding elected her as their federal MP, making her the first elected Green Party Member in Canadian Parliament. In 2012 her fellow MPs voted her "Parliamentarian of the Year." In 2013 they voted her "Hardest Working Parliamentarian of the Year," while in 2014 she earned the title "Best Orator."

Above all these accolades, Elizabeth May is a human being, able to communicate across party lines, beyond politics, meaningfully on a one-to-one basis. She cares to educate, having written eight books. The back cover of her latest book, *Who We Are: Reflections on My Life and Canada*,[601] includes praise from two former prime ministers of Canada, one Liberal and one Conservative. Continually reaching across party lines, she regularly communicates with her constituents, keeping up an informative blog[602] and holding town-hall meetings several times a year across her complex, peninsula- and island-filled federal riding, inviting speakers on finance, the environment and social concerns.

Since the beginning of the Madrona campaign, Elizabeth May has been to each of Madrona Farm's Chef Survival Challenge events. She is traditionally one of the first ticket purchasers, and she is consistently and strongly supportive of the Madrona vision. Nathalie explains why she featured Elizabeth May in the Chef Survival Challenge 2012 movie: "Elizabeth is the most inspiring, intelligent, environmentally wise woman I know. Social justice is of prime importance to her, whether it concerns First Nations, being a woman or being Green. She is always approachable and talks to her constituents like they are human beings. When Elizabeth May got elected, it was another peak experience – just like the success of Madrona's campaign and the moment of Bill Turner's re-election – which made my mind bend. As I see her going into Parliament with love, I would like to see more Elizabeth Mays."[603]

Elizabeth May debates beautifully and never lets an unfair action get overlooked in any little bill. No omnibus has ever gone through Parliament without Elizabeth pointing out its flaws and decrying its passing. As both Nathalie and Robin agree, Elizabeth is absolutely the top-notch parliamentarian any district in Canada has ever seen.

David Suzuki

David Suzuki is another hero high on Nathalie's list.[604] She remembers sitting on her dad's knees and watching *The Nature of Things*,[605] and she has loved receiving Suzuki's teachings ever since. After gaining a Ph.D. in zoology, David Suzuki spent most of his teaching career as a genetics professor at the University of British Columbia. He has published over 50 books, has hosted regular TV and radio educational programs, and has justifiably received more awards than any of us knew existed. As a genetics researcher, he understands clearly why the only benefit of GMOs is to line corporate pocketbooks, and he has been consistent in speaking up about the dangers of genetic manipulation. In return for taking on such huge targets as the GMO and oil and gas industries, he has opened himself up to be bullied at the highest level.

Because corporate powers extend to the media, heroes like Suzuki risk being defamed. The public needs to develop the habit of questioning the source and its motives, instead of the target, especially if the bullying shows up as claims that seem ridiculously exaggerated. As an example of how an ethical hero responds, after one defamatory article in a Canadian newspaper in October 2013, Suzuki stated:

> Some outlets have stooped to ignoring ideas and rational argument in favour of lies, innuendo, exaggeration and personal attacks.
>
> Ironically, one source is a media personality with government ties who has demonstrated a pattern of using bogus arguments and faulty reasoning, leading to a string of libel charges and convictions, censure over violations of the Canadian Broadcast Standards Council ethics code and complaints about racist statements.
>
> It's sad to see so much of our media and governance in such a sorry state that we can't even expect rational discussion of critical issues such as climate change and energy policy. And there is room for debate – not over the existence of climate change or its causes; the science is clear that it is real and that we are a major contributor, mainly through burning fossil fuels and cutting down forests.
>
> But there's room for discussion about ways to address it. And address it we must. We won't get there, though, if we hinder scientists from conducting their research and

speaking freely about it, and if we allow the discussion to be hijacked with petty name-calling and absurd allegations.[606]

Among Suzuki's many gems, he's always stressed the importance of interrelationships. He's pointed out that the small things in our biosphere keep it going for us, but the more we ignore them and pretend they don't matter, the more they're threatened. And so are we. If only we would take the time to simply stand still and observe tiny gifts – preferably in nature – we will really understand how much we can learn from them.[607] He knows, as we all do, that conventional farming 50 or 60 years ago was what we call organic farming today.

David Suzuki knows how to research and how to put that research into easy language for the common person. He contributed an article to the journal *Agriculture and Human Values* in which he quotes two researchers who had reviewed the scientific studies, wondering if we could have both food security and biodiversity. Their review showed that small farms using sustainable techniques were two to four times more energy efficient than large farms. In addition, small farms almost always produced more per hectare than large ones.[608]

As Nathalie understands society's emphasis on money, and clearly sees the monetary value of the trees and the bees, she appreciates David Suzuki's excellence in educating the public about the true costs and the true production value in sustainable farming. But most important for Nathalie, Suzuki constantly holds up the importance of biodiversity in farming. She cherishes his courage, despite the government's attacks, cutting funding for research and muzzling government-paid scientists. True heroes can and do stand on their record, with open books backing up their statements.

HRH, Prince Charles

Another hero who creates a buzz for Nathalie and many others is Prince Charles. She particularly admires him for capitalizing on his royal position to speak openly about the importance of sustainability. In 2012, at Georgetown University in Washington, DC, he gave a speech referring to a UN report on the conclusions of four hundred international scientists, which in summary explains that "[s]mall-scale farmers can double food production within 10 years in critical regions by using ecological methods."[609]

"To feed 9 billion people in 2050, we urgently need to adopt the most efficient farming techniques available," says Olivier De Schutter, UN Special Rapporteur on the right to food and author of the report Prince Charles referred to in his speech. "Today's scientific evidence demonstrates that agro-ecological methods outperform the use of chemical fertilizers in boosting food production where the hungry live – especially in unfavorable environments. . . . We won't solve hunger and stop climate change with industrial farming on large plantations. The solution lies in supporting small-scale farmers' knowledge and experimentation, and in raising incomes of smallholders so as to contribute to rural development."[610]

In his speech, Prince Charles emphasized that small farms using agroecology were the most productive. He said the 2011 report isn't talked about any more, although it certainly should be. He pointed out that the local and organic market in North America and Europe still only represents 2 or 3 per cent of total sales.

Prince Charles has founded an International Sustainability Unit, "to facilitate consensus on how to resolve some of the key environmental challenges facing the world – such as food security, ecosystem resilience and the depletion of Natural

Capital."[611] This sustainability unit has calculated the costs of all the damage from nitrogen fertilizers and other industrial farming practices, clearly showing why these damages need to be factored into any true, total cost comparison between food grown on industrial farms versus food grown on sustainable, organic farms. Governments should make industrial farmers responsible for the costs of pollution clean-up and associated health problems. Also, if subsidies were given as rewards to sustainable farmers instead of to the industrial farmers whose actions produce the damage, then organic, sustainably grown food would become the affordable choice.[612]

Prince Charles has been working on sustainability for over 30 years. A YouTube video of the farm he's saved in the stunning Carpathian Mountains in Romania for The European Nature Trust (TENT) shows how the neighbouring Carpathian farms have remained well-connected with nature.[613] Although Romania has its share of industrial challenges, the farms highlighted in the video demonstrate a continual sustainability going back generations before intensive farming.

We feel hope when a man of such a high position not only acknowledges that the industrial way of farming is destroying the very food base we depend on, but also speaks, writes and acts in order to effect change. "We are going to have to take some very brave steps," he says in his book, *The Prince's Speech: On the Future of Food*. "We will have to develop much more sustainable, or durable forms of food production because the ways we have done things up to now are no longer as viable as they once appeared to be."[614]

Vandana Shiva

Vandana Shiva is another of Nathalie's top heroes. A physicist, ecologist, philosopher, feminist and seed activist/farmer who was born in India to a forest conservationist and farmer mother, she grew up loving nature. After receiving an integrated masters of science honours degree in particle physics from the University of Punjab in 1973, she received her master of arts degree at the University of Guelph, Ontario, in the philosophy of science and her doctorate in philosophy at the University of Western Ontario in 1978. She wrote her dissertation on the philosophical basis of quantum mechanics.[615] Dr. Shiva has the education, experience, honours and energy to counter any provocation she stirs up. She has written over 20 books on a variety of related topics, including but not limited to: saving seeds, water, soil, biodiversity, ecofeminism, sustainability, bio-politics, bio-piracy, patent myths, globalization, democratizing biology and making peace with Earth. Understandably, she's received an equally long list of international awards for her thoughtful, caring accomplishments in preserving our world.

On her website describing the "Rights of Nature," Vandana Shiva writes:

> The Earth's living systems and human communities face multiple crises of climate change, mass species extinction, rampant deforestation, desertification, collapse of fisheries, toxic contamination with tragic consequences for all life. Under the current system of law, Nature is considered an object, a property, giving the property owner the right to destroy ecosystems for financial gain. The Rights of Nature legal doctrine recognizes that ecosystems and plant and animal species cannot simply be objects of property but entities that have the inherent right to exist. People, communities and authorities have the responsibility to guarantee those rights on behalf of Nature. These laws are consistent with indigenous people's concepts of natural law and original instructions as well as the understanding that humans are a part of Nature and only one strand in the web of life.[616]

In April 2010, at the first World People's Conference on Climate Change and the Rights of Mother Earth,[617] held in Cochabamba, Bolivia, she began by describing the "Universal Declaration of Rights of Mother Earth" in which "we are all part of Mother Earth, an indivisible, living community of inter-related and interdependent beings with a common destiny."[618]

In her stance against Monsanto, she speaks eloquently about the drastic side effects of GMOs, about how the company's altered seeds aren't standing up to their creators' claims but are worse than nature's original seeds. Statements she made during Conference 8 as president of The Rights of Nature Tribunal in Quito, Ecuador, on January 17, 2014, include the following:

- "Every aspect of life comes from seed."
- "Food freedom is the freedom of the food."
- "In our cosmology, food is the creator."
- "Industrial monocultures are actually highly unproductive."
- "It is time to act, with love, with deep knowledge and deep resistance."[619]

Dr. Vandana Shiva presented clear closing statements at the same conference. She pointed out that GMOs as a system, with its fossil-fuel energy base, and farms based on industry's way of doing business, are destroying 75 per cent of the planet's soil, water and biodiversity and contributing to 40 per cent of the greenhouse gases. She bluntly labels the GMO system "termina-tor technology" as she feels "it must evolve into a system that can only exist in militarized form, in violence." As she looks at globalization, she sees that instead of democracies provid-ing "representation of the people by the people for the people," they have devolved into "representation of the corporations by the corporations for the corporations." She recognizes that "every destruction is justified in the name of growth" and that "people who lived close to Mother Earth were defined as primitive."[620]

In these same closing remarks, she stated that we need to redefine our categories of "Who is primitive?" versus "Who is advanced?" She pointed out that we must "start recognizing that those who uphold the laws of Mother Earth are the most advanced human beings on the planet – and that the rest of us must follow."[621] We need to start recognizing not only a science of interconnectedness but also the fragility of that interconnectedness when we start interfering with ecosystems. We need to recognize the culture of fear being built by those who attack the defenders of Earth. In that moment, we must instead honour those who protect nature's seeds and pollinators, which have been succeeding for thousands of years according to the laws of Earth – not the laws of Monsanto. Dr. Shiva also encouraged all Indigenous peoples to say "no" to colonization.

Her free e-book *The GMO Emperor Has No Clothes* clearly explains what is going on in the GMO system. Of course, the GMO Emperor is Monsanto, with its economy-destroying seeds, its Roundup and other sprays and fertilizers that make plants develop worse diseases than those they were already suffering from. Shiva also writes about the ocean's huge dead zones, and why thousands of bees have been killed off, almost to the point of extinction.[622]

Sophie Wooding's cousin has had the good luck to go to Vandana Shiva's farm in India. Called the Navdanya Trust, its purpose is to train farmers. While he was there, he heard Shiva speak about sovereignty and democracy. She believes that Earth has a right to democracy as much as the people do. Together, people and Earth are sovereign. She bases her work on five sovereignties: seeds, food, land, water and forest. For example, rivers are sovereign to flow and people are sovereign to collect water, as long as the river's limits are respected.

Shiva points out that arrogance builds up when greed and stupidity get together. For example, in a recent campaign, Shiva drew an audience's attention to the fallacies in the claims made in 2013 by producers of GM bananas. The banana is certainly a good food, and India is its largest producer. In India, iron deficiency has been on the rise over the last 30 years. From the statements made by GM banana producers, consumers could be beguiled into thinking that bananas must have a lot of iron. But there is much irony (pardon the pun) in this little story.

In fact, iron is much more abundant in the voluntary greens with the common English name lamb's quarters (*Chenopodium album*). Called *bathua* in India, the greens of this plant taste quite a bit like spinach, but they have triple the calcium, more than double the Vitamin C, more Vitamin A and even half the iron of spinach.[623] Some people would call them weeds, but most people in India grew up eating those greens in many different dishes. Even people who had no land could pick these "weeds," and as long as they ate them, they were not iron deficient. Other iron-rich items that grow wild in India include mangoes, the drumstick tree (*Moringa oleifera*) (which produces long seed pods with seeds that have a horseradish-like taste), and cumin (*Cuminum cyminum*).

When industrial farmers in India started using chemical fertilizers and spraying herbicides, they wiped out most of these healthy weeds. The genetic engineers say they're going to engineer an "iron-rich banana." Even if they double or triple the amount of iron in these bananas, it wouldn't match the amount in lamb's quarters, which would still have seven times as much iron. Of course, that doesn't even touch the fact that lamb's quarters, compared with bananas, has almost two hundred times as much Vitamin A, more than eight times as much Vitamin C and many other benefits.

This story about bananas and lamb's quarters illustrates Vandana Shiva's phrase, "the monoculture of the mind," which she used in her acceptance speech on December 9, 1993, for The Right Livelihood Award (given for "outstanding vision and work on behalf of our planet and its people"). "The monoculture of the mind," Shiva said, "treats all diversity as disease, and creates coercive structures to model this biologically and culturally diverse world of ours on the privileged categories and concepts of one class, one race and one gender of a single species."[624]

This kind of thinking (or non-thinking) makes people blind to the wealth of nature's diversity. Counting on even further public ignorance, corporations have had the gall to call their GMO and pesticide products vanguards of the "Green Revolution!"[625] Thus, Dr. Shiva persists in educating as many people as she can, pointing out that Monsanto has a huge concentration of power, but it can't even begin to do a competent job of creating good-quality food. She reiterates that the land should not be for sale because it is sovereign, which explains why she set up Navdanya Trust in 1984, when she saw how devastating the loss of biodiversity was in India. Over the last decade and a half in India, more than 250,000 farmers committed suicide after being trapped in hideous debt cycles brought on by the multinational seed and chemical corporations.[626] As this equates to 17,000 farmers committing suicide every year, compassion has driven Dr. Shiva to educate farmers and others.

Now farmers at Navdanya are reducing their expenses by the same 90 per cent that chemical corporations charged to cover their chemicals and profits. As a result, Navdanya farmers earn three times more than chemically dependent farmers.[627] Over the past 20 years, Dr. Shiva's Navdanya Trust has given seeds to farmers who need them, whether due to debt or natural disasters. Through hands-on experience, farmers also receive an opportunity to become educated about seed sovereignty and sustainable agriculture.[628]

Let's All Save Farmland Forever

It's vital that we choose heroes who mean the most to each one of us. Nathalie feels she witnessed a miracle when Madrona Farm was finally protected. When she stood on the logging roads, the world's problems hung heavily over her heart, and nothing seemed to work. When she applied her heroes' lessons of balancing power and love, 3,500 people pulled together to protect Madrona Farm. She realized she'd found a pathway out of distress.

As we discover how people – and our heroes – succeed, we're inspired to decide how we can apply their gems to our own processes moving forward. We can make the choice to stick with hope and work together.

By practising agroecology, we can rebuild our broken bonds with nature. By honouring biodiversity in all its resilience, we can mitigate climate change. We hand these ideas, concepts and principles to the commons to be used for our common, sustainable welfare.

We welcome our readers to choose to make a difference, for ourselves and for the world.

ACKNOWLEDGEMENTS

Nathalie's Acknowledgements

I would like to extend a big thank you to Robin and Sophie. If it weren't for them, this important information would not have been documented.

Special thanks to David Chambers, the love of my life. And to my family: Georgette McBain, John Viola and Allen McBain. Also to my children, Sage Rogers and Lola Chambers, for putting up with us.

Thanks to Bruce Chambers for his generosity, which made it all possible. To Russell Chambers for his patience and Derek Chambers for challenging and allowing us to have this great opportunity for a community movement.

Thanks to Ruth Chambers – the most amazing conservationist I never met – through whose works I became inspired.

To Ed and Lynne Johnson, Barbara Souther and the Farmlands Trust – this project would not have been possible without you. Mel McDonald – thanks for listening to your angels!

Thanks also to the 3,500 people who donated to Protect Madrona Farm Forever. There were literally thousands who supported us in a multitude of ways during "the years of ideas" stage of campaigning, educating and then fundraising. We acknowledge you – including all the chefs of Chef Survival Challenge, past, present and future – and thank you for believing.

You all know who you are. The following list is not exhaustive, but it gives some idea of our incredibly numerous supporters, to whom we are eternally grateful. If we have forgotten anyone, please forgive us.

Dawn Meredith, Steve, Joan Mann and the whole Mann clan, Margo, Gordon and Joan Alston Steward, Dawn and Barry Loucks, David, Jennifer and Amelia Scoones, Robert and Linda (Still Life), Jenny Whitfield, Jonathon/Tessa, Joan and Lorna, John and Renata Ferguson, Jack and Dallas, Sharon Jack, Laura-Leigh, Ramona Scott, Paula Hesje, Matt Phillips, Carla Funk, Tom Arnold, Tim Maloney, Connie, Jennifer Hobson, Heather Skyte, Cassie, Jill Patterson, Lorelle Polsten, Carmen, Torrance Coste, Bill Turner, Wendy Innes, Andrew and Nick, Heather Skype, Alison Spriggs, Nichola Walkden, Ian Fawcett, Briony Penn, Bob Ray, Robin Roberts, Diana Denny, Anita Harwood, Phyllis Mercer, Ray Galey, Bob Galey, Steve Fisher, Linda Cirrella, Terresa, Jasper, Yanni and Pernilla, Joan and Michel Motek, Ted and Lora Lea, Peter Muse and family, Jim Cassels and the Cassels family, Jim Hutchison, Kathleen Sheppard, Lorella Polsten, John Shields, Dave Cutler, Chris Loren, Jeremy Baker, Judy Maskulak, Colleen Maskulak (Kolleen Heal), Kim Harrison, Erin Cooper, Claudia, Jason Leblanc, Heather Coey and her dad, Ken Agate (the Blethering Place) Sean O'Keefe, Darren and Claudia Copely, former Saanich mayor Frank Leonard, Susan Brice, Vic Derman, Dean Murdock, Judy Brownoff, Vicki Saunders, Ben Isitt, Melissa Moraz, Aviva Isitt, Lorna Knowles and Bill Finley from Hemp & Company, Don Eastman, Francis, Ellie and Lars, Trevor Lantz, Dr. Nancy Turner, Lana Popham, Karen Platt,

Chris Denny, Bob Maxwell, Ryan Vantreight, Guy Dauncey, Carolyn Harriot, Erin, Mark and Sophia, Don, Carolyn and Aidan, Knight, Mae Moore, Lester Quitzau, David Mincey, Page Roberts, Ken Kelly, Gregor Roberts, Ken Houston, Ken Harper, Dan Hayes (London Chef), Chris Hammer (Royal Colwood Golf Course), John Brooks, Rob Cassels (Pablo's Dining Lounge), Rob Ringma, Tanya Smith, Dana Hutchings, Johnny Marcolon, Nev Gibson, Marcella, George Harrison and the Gulf Islands Film and Television School (GIFTS), Vancouver Island Helicopters, Adrian Lam, Judith Lavoie, Grania Litwin, Margo Aker, Darlene and Bill Hayes, Phyllis and Gordon, Bill and Judy Gaylord, Daphne Goode and Lorrain Scullen at Shaw TV (and Michael Wylie, Shane Yakelashek and Karen Elgersma formerly of Shaw TV), Andrew Ainsley at B Channel News, Cathleen Jardin, Gary Hynes, Greg Aspa, Joe Blake, Helen and Glenn Sawyer, Alan and Susan Walton, Bea and Pat Summers, John Brooks (Spinnakers), Stephan Drolct (Camille's), Harold, Daphne and Fraser Wolf, Garry and Darlene McCue, Ruth and Allan Holmes, Roy, Joy, Dorothy and Mark Hawes, James Miskelly, Julian Crown, Sean Cabers (Carl), Luke Woodyard, Jason Slade, Gordon Jarvis, Sophie Wooding, Beth Supple, Gordon Ruby, Shawn Green, Vince Vaccaro, Leroy Stagger, Eric Raschig and The Turnpike Bandits, Raven Hill Farm, Tony Young, Jeff Campbell, Sylvia Myren, the "White Knight," Ed and Kathy Lowe, Nora Wall and family, the "Two Tanya Team" and Cathy Baan (formerly of the *Times Colonist*), Sharon Tiffon at the *Saanich News*, Sharon Jack, Brad Williams, Pat McGuire, Pat, Michael, Matt Martin of Capital Iron, Sandra Richards, the Victoria Foundation, Mike and Dianne Brooks, Jerry and Julia Kruz, Sunny and Josee, Nelly Furtado, Mike Dunlop, Garett Schack (Vista 18), Solomon Segal of Pagliacci's, Cory Pelan of The Whole Beast (and Cory's mom, wife, Madonne, and daughter, Paris), Eden Oliver, David and Page Mincey, Heather Hammer, Ray Parkes, Cliff, Jiles Barett, Dr. Eric Higgs, James Lawson and family, Flora Kyle-Wood, Linda Geggie, Ray and Muriel Cooper, Karen and Bert Harrison, Leif Wergeland, Paul Gerrard, Laura Anderson, Andrea Ray, Seth Wright, Ron Carter, Craig Sorochan, Bill McKechnie, Stu Nempkin, Sally Glover, Ryan and Kristy Taylor at Café Fantastico and its crew of "Fantasticats" (Stu Garret and Ariella and Gareth), Karen Chester, Lana Popham, David Cubberley, Carol Pickup, Patrick Stewart, Robert Bateman, Jack and Dallas, Sarah Pugh, Cosmo Means, Ron Carter, Mike Dunlop (Vista 18), Noel Richardson and Andrew Yeoman (formerly of Ravenhill Farm [Mount Newton]), Sean Brennan, Lorna and "Queen E" Joan, Fran Pugh, Alistair Craighead, Christine Torgensen, Thomas Loo, Susan Tychie and Lee Fuge, Ted and Laura Lea, Ted and Josie Newman.

Robin's Acknowledgements

I really appreciate Nathalie and Dave's deep connection to the earth, their ongoing desire to respect the ecosystem at every level, and their incredible passion for organic, sustainable farming. Above all, I'm so glad they insisted on sharing it so enthusiastically. My family and I have savoured the taste of Madrona Farm's fresh, sweet produce for more than a decade. As Dave and Nathalie educated us at their food stand – both in words and with incomparable flavours – I began to realize that I'd like to encourage as many people as possible, around the world, to reach for similar food. As the first step entails saving farmland, how could I resist Nathalie's request to co-author *Saving Farmland*? With Sophie as an intelligent co-author who consistently communicated with sensitive insight, I felt especially well-supported.

As a person who loves learning, I relish the fact that the process of writing this book taught me a huge amount about respecting the complex inter-relationships involved in growing healthy food for the world. I am grateful to both Arran Stephens and Harold Steves for sharing their family stories of farming – especially for the different, inspiring ways they encourage others to save farmland in a sustainable manner. I also appreciate Bill Turner for his vast experience, for his incredible ethics through difficult times and for guiding us through both land trust and political issues.

I am grateful for Meaghan Craven's frank and scientifically questioning mind, providing editorial insight to help me delve even more deeply into researching details and considering stylistic particulars that provided ongoing growth opportunities. Proofreading editor Sarah Weber's clarity in catching other niggling details helped us move efficiently through the final stages. Throughout our project, I felt relieved by the perceptive, forward-moving support offered by Rocky Mountain Books publisher Don Gorman.

Endless hugs go to my practical, caring, hardworking wife, Diana Denny, for taking over many of the duties we customarily share together while this book blossomed. To our many friends and family members who kindly offered their curiosity and encouragement – or simply an attentive ear – thank you! All this support truly shows me how a caring community sustains us all.

Sophie's Acknowledgements

First of all, I'd like to thank Rocky Mountain Books and particularly Don Gorman, for having the patience to see this project through, even when witnessing our collaborative process could have easily provided them with enough doubt to quit. What a supportive team! Thank you to our editor, Meaghan Craven, as well, for having such a good eye for consistency, among many other editing and creative talents.

Thanks to Robin for always, always seeking a positive way to say things, but also being my mutual rant partner over the past couple years. To Dave, Nathalie and Lola, thank you for treating me like family. Thank you for cooking me gourmet meals when I'm there *not* as an employee, and for always sending me home with delicious produce. As well, thank you to all my friends in the agroecological world: those who let me volunteer in their gardens and those who took on a permaculture design course with me and matched my excitement about studying soil and digging up the roots of what's practical for us to employ in our lives, right now.

To Gordon, thank you so much for taking me to Madrona in the first place. You introduced me to all this craziness, and even when I've thought the whole thing was a bite or two more than I could handle, I've always been glad that you brought me into this mess with such passion.

Thanks to my writing mentors and peers, particularly at University of Victoria, for challenging me to keep writing post-graduation. Thanks to my friends and family (Mom, Dad, Kelsy, Andrew, Erika, Isaiah, Elliott, Faith) for letting me talk about this project whenever I needed to. Thanks to everyone who asked me questions, read partial (sometimes painful) first drafts and acted excited about reading the book as soon as it comes out. Among my friends, I want to say a special thanks to Aaron – my running partner, romantic interest and a fellow writer – for understanding and allowing me to beat myself up when that's what I really wanted to do, and for laughing at my ridiculous antics when my brain was fried.

And to all the rest of you who made this project possible through both your stories and your readership – those I've met and those I haven't had the pleasure of meeting – a heartfelt thank you!

SELECT BIBLIOGRAPHY

We provide the following resources to help whet your appetite and fuel optimism for healthy farmland. For further resources, take a look at the endnotes to the individual chapters.

Books and Articles

Benyus, Janine. *Biomimicry: Innovation Inspired by Nature*. New York: William Morrow, 2002.

Cobb, Tanya Denckla. *Reclaiming Our Food: How the Grassroots Food Movement Is Changing the Way We Eat*. North Adams, Mass.: Storey Publishing, 2011.

Davis, Wade. *The Wayfinders: Why Ancient Wisdom Matters in the Modern World*. Toronto: House of Anansi Press, 2011.

Deur, Douglas, and Nancy Turner, editors. *Keeping It Living: Traditions of Plant Use and Cultivation on the Northwest Coast of North America*. Vancouver: University of British Columbia Press, 2006.

Druker, Steven M. *Altered Genes, Twisted Truth: How the Venture to Genetically Engineer Our Food Has Subverted Science, Corrupted Government, and Systematically Deceived the Public*. Salt Lake City, Utah: Clear River Press, 2015.

Estabrook, Barry. *Tomatoland: How Modern Industrial Agriculture Destroyed Our Most Alluring Fruit*. Kansas City, Mo.: Andrews McMeel Publishing, 2012.

Fukuoka, Masanobu. *The One-Straw* Revolution. New York: Rodale Press, 1978.

Hesterman, Oran B. *Fair Food: Growing a Healthy, Sustainable Food System for All*. New York: PublicAffairs, 2012.

HRH, The Prince of Wales. *The Prince's Speech: On the Future of Food*. Emmaus: Rodale Books, 2012.

Imhoff, Daniel. *The CAFO Reader: The Tragedy of Industrial Animal Factories*. Healdsburg, Calif: Watershed Media, 2010.

Johnson, Lorraine. *Tending the Earth: A Gardener's Manifesto*. New York: Viking, 2002.

Kingsolver, Barbara, with Steven L. Hopp and Camille Kingsolver. *Animal, Vegetable, Miracle: A Year of Food Life*. New York: HarperCollins, 2007.

Mader, Eric, Matthew Shepherd, Mace Vaughan, Scott Black, with Gretchen LeBuhn. *Attracting Native Pollinators: Protecting North America's Bees and Butterflies*. North Adams, Mass.: Xerces Society/ Storey Publishing, 2011.

Moore Lappé, Frances. *EcoMind, Changing the Way We Think, to Create the World We Want*. New York: Nation Books, 2011.

———. *Hope's Edge: The Next Diet for a Small Planet*. New York: Putnam, 2002.

Muir, John, with photographs by Scot Miller. *My First Summer in the Sierra, Illustrated Edition*. New York: Houghton Mifflin Harcourt, 2011.

———. *Our National Parks*. Sierra Club, accessed June 16, 2014, http://vault.sierraclub.org/john_muir_exhibit/writings/our_national_parks/.

Pawlick, Thomas. *The End of Food: How the Food Industry Is Destroying Our Food Supply, and What You Can Do About It*. Vancouver: Greystone Books, 2006.

Rabhi, Pierre. *As in the Heart, So in the Earth: Reversing the Desertification of the Soul and the Soil*. Translated by Joseph Rowe. New York: Park Street Press, 2006.

Suzuki, David. "Food and Climate Change." *Food and Our Planet* (blog). Accessed March 16, 2015. http://www.davidsuzuki.org/

what-you-can-do/food-and-our-planet/food-and-climate-change/
?gclid=CMSVpbLa3bwCFYdgfgodiXEACQ.

———. "Thought for Food: Organic Farming Is Good for You and
the Planet." *Docs Talk* (blog). October 7, 2010. http://www.
davidsuzuki.org/blogs/docs-talk/2010/10/thought-for-food-
organic-farming-is-good-for-you-and-the-planet/.

———. "We Can Learn from Nature's Genius." *Science Matters*
(blog). November 1, 2012. http://www.davidsuzuki.org/blogs/
science-matters/2012/11/we-can-learn-from-natures-genius/.

Suzuki, David, and Ian Hanington. *Everything Under the Sun:
Toward a Brighter Future on a Small, Blue Planet.* Vancouver:
Greystone Books/David Suzuki Foundation, 2012.

Suzuki, David, with Amanda McConnell and Adrienne Mason. *The
Sacred Balance: Rediscovering Our Place in Nature.* Vancouver:
Greystone Books/David Suzuki Foundation, 2007.

Turner, Nancy. *Ancient Pathways, Ancestral Knowledge: Ethnobotany
and Ecological Wisdom of Indigenous Peoples of Northwestern
North America.* Montreal/Kingston: McGill-Queen's University
Press, 2014.

———. *Food Plants of Coastal First People.* Victoria: Royal British
Columbia Museum, 1995.

———. *The Earth's Blanket: Traditional Teaching for Sustainable
Living.* Vancouver: Douglas & McIntyre, 2005.

Weber, Karl. *Food Inc.: How Industrial Food Is Making Us Sicker,
Fatter, and Poorer – and What You Can Do about It.* Los Angeles
and New York: Participant Media and PublicAffairs Books, 2009.

Websites Worth Visiting

Janine Benyus, Biomimicry: http://biomimicry.net/.

Life Cycles Project, Cultivating Communities: http://
lifecyclesproject.ca/.

Open Permaculture School, Regenerative Leadership Institute:
https://www.openpermaculture.com/.

"Rewilding Vancouver – Fall 2013." 16 YouTube videos. Posted by
"DavidSuzukiFDN," last updated April 16, 2014. http://www.

youtube.com/playlist?list=PLK1TK6eY3GAj37heGC_vqYJFR8KZ_
1rvS.

"Rewilding Vancouver – UBC Farm." The University of British
Columbia. March 23, 2014. http://ubcfarm.ubc.ca/2014/03/23/
rewilding-vancouver-ubc-farm/.

Satish Kumar's "ideas globe" (29 webpages full of proposals, projects
and actions generated by the community, including resources),
Our Future Planet: http://www.ourfutureplanet.org/ideas-globe/
wheel.

WWOOF, World Wide Opportunities on Organic Farms: http://
www.wwoof.net/.

NOTES

2 "Apple as Planet Earth," You-
Tube Video, 1:02, posted by
"American Farmland," December
14, 2009, http://www.youtube.
com/watch?v=_J9cg7dxD5E. We
urge you to watch this video by
American Farmland Trust (also
available at: http://www.farmland.
org/images/flash/apple.swf).
You'll come away anxious to act.

3 John Vidal, "Land Acquired
over Past Decade Could have
Produced Food for a Billion
People: Oxfam Calls on World
Bank to Stop Backing Foreign
Investors Who Acquire Land
for Biofuels That could Produce

Food," *The Guardian*, October 4,
2012, http://www.guardian.co.uk/
global-development/2012/oct/04/
land-deals-preventing-food-
production.

4 "Farmland by the Numbers: The
National Resources Inventory,"
American Farmland Trust,
accessed June 5, 2014, http://
www.farmland.org/programs/
protection/American-Farmland-
Trust-Farmland-Protection-
Farmland-by-the-numbers.asp.

5 "Report of the Special Rapporteur
on the Right to Food, Olivier De
Schutter, Addendum, Mission
to Canada," United Nations
General Assembly, December
24, 2012, 4, item 6, http://www.
srfood.org/images/stories/
pdf/officialreports/20121224_
canadafinal_en.pdf.

6 "Resources," Planted Network,
accessed June 5, 2014, http://
www.plantednetwork.ca/about/
about.aspx?page=4&gclid=CP7y
tp2SrbgCFQ9eQgodLRsA8g. For
more details on BC food costs,
see "Cost of Eating in British

Columbia 2011," Dietitians, accessed September 21, 2014, 6–8, http://www.dietitians.ca/ Downloadable-Content/Public/ CostofEatingBC2011_FINAL. aspx.

7 Herb Barbolay, Vijay Cuddeford, Fern Jeffries, Holly Korstad, Susan Kurbis, Sandra Mark, Christiana Miewald and Frank Moreland, "Vancouver Food System Assessment" (Vancouver: Western Economic Diversification Canada, the City of Vancouver Department of Social Planning, Simon Fraser University Centre for Sustainable Community Development, Environmental Youth Alliance, 2005), http://www.sfu.ca/ content/dam/sfu/cscd/PDFs/ researchprojects_food_security_ vancouver_food_assessment%20 (short).pdf.

8 Lynn McIntyre, Gordon Walsh and Sarah K. Connor, "A Follow-up Study of Child Hunger in Canada," Working Paper W-01-1-2E, Applied Research Branch, Strategic Policy (Ottawa: Human Resources Development Canada, 2001), 5, http:// publications.gc.ca/collections/ Collection/MP32-28-01-1-2E.pdf.

9 Ibid., 3.

10 Laura Payton, "UN Official Sparks Debate over Canadian Food Security," *CBC News*, May 16, 2012, http://www.cbc.ca/ news/politics/story/2012/05/16/ pol-un-canada-food-security. html. Also see Food Secure Canada's newsletter at http:// foodsecurecanada.org/ fsc-newsletter-2012-06.

11 *Loss and Fragmentation of Farmland* (Edmonton: Resource Planning Group, Policy Secretariat, Alberta Agriculture, Food and Rural Development, 2002), I, ii, 3, and 83, http://www1.agric. gov.ab.ca/$department/ deptdocs.nsf/all/psc4786/$file/ loss_fragmentation. pdf?OpenElement.

12 The World Bank, "Arable Land (% of land area) in Canada," Trading Economics, accessed September 21, 2014, http:// www.tradingeconomics.com/ canada/arable-land-percent- of-land-area-wb-data.html. See also "Arable Land (% of land area)," The World Bank, accessed September 21, 2014, http://data.worldbank.org/ indicator/AG.LND.ARBL.ZS and http://wdi.worldbank.org/ table/3.1, where percentages of arable land, forest land, permanent cropland, land use and rural population numbers are provided for many different countries.

13 See chapter 14 of this book to find out more about the "Green Revolution."

14 Peter Rosset, "Lessons from the Green Revolution," March/April 2000, Iowa State University, Department of Sociology, accessed September 22, 2014, http://www.soc.iastate.edu/ sapp/greenrevolution.pdf. This article also can be found on the Food First website (http:// www.foodfirst.org). Food First, the Institute for Food and Development Policy, was founded by Joseph Collins and Frances Moore Lappé in 1975. "Lessons from the Green Revolution" is based on research

presented in Frances Moore Lappé, Joseph Collins and Peter Rosset, with Luis Esparza, *World Hunger: 12 Myths* (New York: Grove Press/Earthscan, 1998).

PART ONE: UNDERSTAND THE ISSUES

One: Hear the Stories

15 Harry S. Truman, "128. Statement by the President upon Signing the National School Lunch Act," June 4, 1946, accessed June 7, 2014, https://www.trumanlibrary. org/publicpapers/index. php?pid=1570&st=&st1=.

16 Unless otherwise noted, quotes from Arran Stephens in this chapter come from a version of his story adapted by the authors, and used with his permission, per e-mail correspondence between Robin Alys Roberts and Arran Stephens, December 4, 2013.

17 For more information on deer, see British Columbia Ministry of Environment, Lands and Parks, "Mule and Black-Tailed Deer in British Columbia: Ecology, Conservation and Management," accessed June 5, 2014, http://www.env.gov. bc.ca/wld/documents/muledeer. pdf; and "Deer Signs," The Deer Initiative, accessed June 5, 2014, http://www.thedeerinitiative. co.uk/uploads/guides/170.pdf.

18 This can be freely downloaded at http://us.naturespath. com/blog/2009/07/10/ sawdust-my-slave.

19 From Arran Stephens, "Humble Beginnings,"
Nature's Path, March 28, 2009, http://ca-en.naturespath. com/blog/2009/03/28/ humble-beginnings.

20 Ibid.

21 Thomas Kidd, *History of Lulu Island and Occasional Poems* (Wrigley Printing Company, 1927; Richmond, BC: City of Richmond Archives, 2007), 18, http://www.richmond.ca/__ shared/assets/History_of_Lulu_ Island_by_Thomas_Kidd18191. pdf.

22 Harold L. Steves Sr., January 17, 1973, from page 2 of notes typed by Dellis Cleland, accessed September 22, 2014, http://www. richmond.ca/__shared/assets/ OH-Steves_Sr36247.pdf.

23 "Steveston," Wikipedia, accessed June 10, 2014, http:// en.wikipedia.org/wiki/ Steveston,_British_Columbia.

24 Harold Steves Sr. interview by David Jelliffe, February 7, 1972, Tapes 6 and 7. See the full transcription (with "no restrictions") at http://www. richmond.ca/cityhall/archives/ search/histories/stev.htm.

25 Kidd, *History,* 40.

26 Dellis Cleland, "Fog and Bog and Peat Fires," Department of Geography, University of British Columbia, accessed September 23, 2014, http://ibis.geog.ubc.ca/ richmond/city/fire.htm. This version is reprinted from the original paper by Dellis Cleland (1972) prepared for the Richmond Nature Park. It is based on over one hundred oral interviews compiled between February and June 1972, under a Federal Local Initiatives project

grant, for the Richmond Nature Park Committee.

27 Richard Steven Street, *Beasts of the Field: A Narrative History of California Farmworkers, 1769–1913* (Redwood City: Stanford, 2004), 266, accessed September 23, 2014, http://sanjoaquinhistory.org/blog/?p=160.

28 Cleland, "Fog and Bog."

29 Ibid.

30 Norris Adams, "The Late Kitsilano Railway Trestle, 1886–1982," ExpoRail, accessed September 24, 2014, http://www.exporail.org/can_rail/Canadian%20Rail_n0373_1983.pdf.

31 Although the Steveses called this disease "granulism," it seems related to the disease called "glanders," caused by animals ingesting contaminated food or water. Lesions form in the lungs and ulcerate mucous membranes in the upper respiratory tract. The cells enlarge, decay, granulate and become infectious. See "Glanders," Wikipedia, accessed June 12, 2014, http://en.wikipedia.org/wiki/Glanders.

32 "Dr. Lee's Interview with Harold Steves," Brio Integrative Health Centre, Inc., June 7, 2014, http://www.yourbriohealth.com/news/dr-lees-interview-harold-steves.

33 Rod Mickleburgh, "Harold Steves' Unwavering Passion for the Land," *The Globe and Mail*, August 19, 2013, http://www.theglobeandmail.com/news/british-columbia/harold-steves-unwavering-passion-for-the-land/article13835932/.

34 Douglas Deur and Nancy K. Turner, eds., *Keeping It Living* (Vancouver: University of British Columbia Press, 2006).

35 See Hank Shaw, "Blue Camas and Other Edible Bulbs," Hunter-Angler-Gardener-Cook, accessed June 6, 2014, http://honest-food.net/2011/07/21/blue-camas-and-other-edible-bulbs/.

36 Dana Lepofsky, Douglas Hallett, Ken Lertzman, Rolf Mathewes, Albert (Sonny) McHalsie and Kevin Washbrook, "Documenting Precontact Plant Management on the Northwest Coast: An Example of Prescribed Burning in the Central and Upper Fraser Valley," in *Keeping It Living*, ed. Douglas Deur and Nancy J. Turner (Vancouver: University of British Columbia Press, 2006), 238–39.

37 Daniel Green, "Reviving an Ancient Agricultural Practice: The Root Gardens of Canada's West Coast Aboriginals," *Soiled & Seeded: Cultivating a Garden Culture* 9 (Autumn 2013), http://soiledandseeded.com/magazine/issue06/root_gardens.php.

38 Listen to Nancy J. Turner's archived webcast, "Seasons and Biocultural Diversity in a Changing World: Just When the Wild Roses Bloom," Canadian Museum of Nature, accessed June 7, 2014, http://nature.ca/en/explore-nature/blogs-videos-more/webcasts/seasons-biocultural-diversity-climate-change.

39 E. Richard Atleo, "Preface," in *Keeping It Living*, ed. Douglas Deur and Nancy J. Turner (Vancouver: University of British Columbia Press, 2006), x.

40 Ibid.

41 "YOS," First Nations: Land Rights and Environmentalism in British Columbia, accessed June 9, 2014, http://www.firstnations.de/development/coast_salish-yos.htm.

42 Wilson Duff, "The Fort Victoria Treaties" *BC Studies* 3 (Fall 1969), http://prophet.library.ubc.ca/ojs/index.php/bcstudies/article/download/607/651. On page 25, Duff writes: "The payments, evidently, were made in the form of blankets. Douglas reported that he paid the Saanich 'in woolen goods which they preferred to money.'" On the next page, Duff reported: "The Saanich treaties leave us with a number of puzzles. The 50 blankets noted as paid to the South Saanich, at 16/8 each, do indeed total the £41 .. 13 .. 4 on the face of the treaty. But only ten men are listed, and 5 blankets each would seem more than their share. Later, 117 more men were paid. The total cost to the Company was £109 .. 7 ..6. That amount was worth 386 blankets (at 5/8) plus two pence left over (another arithmetical error?). That number of blankets would be 3 each for the 127 men who made their marks, with 5 left over; or 3 each for the 128 men Douglas thought he had paid, with 2 left over. But by now we have reached the realm of pure speculation, and have run out of clues."

43 Ibid.

44 "YOS."

45 Edward Hill, "First Nations Act to Reclaim Name of Mount Doug," *Victoria News*, May 18, 2013, http://www.vicnews.com/news/207936901.html.

46 Vancouver Island has 18 mountain ranges.

47 "Average Annual Precipitation for British Columbia," Current Results, accessed June 8, 2014, http://www.currentresults.com/Weather/Canada/British-Columbia/precipitation-annual-average.php.

48 See "Max Enke Fonds," Memory BC, accessed June 10, 2014, http://www.memorybc.ca/max-enke-fonds;rad.

49 "Bluffs Park," The Galiano Club, accessed June 10, 2014, http://galianofoodprogram.ca/bluffs-park.

50 Visit the society's website at: http://www.vicnhs.bc.ca/.

51 "Blenkinsop Farmer Was True Role Model," *Times Colonist*, November 30, 2007, http://www.canada.com/story_print.html?id=b95af53f-b18f-4396-9880-e90b73d9f05e&sponsor.

52 Leslie Puska, Liz Clements and Kelsey Chandler, "Climate Change and Food Security on Vancouver Island: Discussion Paper," Office of Community Based Research, August 2011, 7, http://www.uvic.ca/research/centres/cue/assets/docs/Climate%20Change%20and%20Food%20Report.pdf. "Fifty years ago, the island produced 85% of its own food. Today, island farms produce about 5% of all of the food consumed

(MacNair, 2004). The rest is imported onto the island from the mainland, from elsewhere in Canada, and from other countries." See also page 9 of this paper: "Vancouver Island's available food supply is further compromised due to the Island's loss of its food processing and storage facilities (Stovel, 2008). Having these facilities would enable farmers to store the surplus from their summer and fall harvests, which could be drawn from as needed during less productive seasons. Instead, much of the Island's seasonal surplus either goes to waste or is exported. Because of this lack of storage and processing facilities to secure the food produced locally, available stocks are predicted to only sustain the population for two to three days in the winter season (Mark, Moorland and Gage 2007)."

53 Masanobu Fukuoka, *The One-Straw Revolution: An Introduction to Natural Farming,* trans. and ed. Larry Korn (New York: New York Review Books, 2009). This edition of the book has a wonderful introduction by Frances Moore Lappé. As well, consider visiting YouTube to watch the book's nine-minute trailer, edited by editor and translator Larry Korn: "The One Straw Revolution by Masanobu Fukuoka," YouTube video, 9:08, posted by "5obeyond," September 21, 2010, http://www.youtube.com/watch?v=XSKSxLHMv9k.

54 Masanobu Fukuoka, *The One-Straw Revolution* (New York: Rodale Press, 1978), 119.

55 For all-encompassing information on Fukuokoa's methods, see Anthony Delfau, "On Seedballs," On Seedballs, accessed June 8, 2014, https://sites.google.com/site/onseedballs/. And, for restoration projects worldwide, see Anthony Delfau, "Restoration Projects Worldwide," On Seedballs, accessed June 8, 2014, https://sites.google.com/site/onseedballs/worldwide-projects.

56 For a map of farmland in the ALR in Saanich, showing the property inside and outside the Urban Containment Boundary (including Gordon Head) as of September 2013, see http://saanich.ca/living/afs/docs/farm_map_jul_2011.pdf. "380 farms received 2010 farm classification status for tax purposes." Of the 10,733 hectares (21,522 ac) in Saanich, 1,844 hectares (4,557 ac) (or 17 per cent) had ALR status. See also District of Saanich, "About Agriculture & Food Security, *Agricultural Areas*," Saanich, accessed December 19, 2014, http://saanich.ca/living/afs/agricultureabout.html.

Two: Welcome the Bees

57 Sue Monk Kidd, *The Secret Life of Bees* (New York: Penguin, 2002), 92.

58 For an informative article with several photos of bees pollinating flowers – and six photos of a bee lifting its leg(s) when someone strokes it – see Beatriz Moisset, "Bumble Bees: Panda Bears of the Insect World," Native

Plants and Wildlife Gardens, accessed June 10, 2014, http://nativeplantwildlifegarden.com/bumble-bees/.

59 Eric Mader, Matthew Shepherd, Mace Vaughan and Scott Black, with Gretchen LeBuhn, *Attracting Native Pollinators: Protecting North America's Bees and Butterflies, The Xerces Society Guide* (North Adams, Mass.: Storey Publishing, 2011), 26.

60 W.H. Davies, "Leisure," in *Songs of Joy and Others* (London: A.C. Fifield, 1911).

61 See the society's website: http://www.xerces.org/.

62 To listen to buzz pollination, watch "Buzz pollination Senna bumblebee 8 2013 w," YouTube video, 0:41, posted by "Beatriz Moisset," https://www.youtube.com/watch?v=YP3lofqejfA. See also, Mader et al., *Attracting Native Pollinators*, 45.

63 "Organic Pollinator Habitat Assessment," The Xerces Society, accessed June 10, 2014, http://www.xerces.org/wp-content/uploads/2009/11/OrganicPollinatorHabitatAssessment.pdf.

64 For photos of some of the different bumblebees in southern BC, see "Bumble Bees," Bee Friendly, accessed June 10, 2014, http://beefriendly.ca/bumble-bees/.

65 Marge Dwyer, "Study Strengthens Link between Neonicotinoids and Collapse of Honey Bee Colonies," Harvard School of Public Health, May 9, 2014, http://www.hsph.harvard.edu/news/press-releases/study-strengthens-link-between-neonicotinoids-and-collapse-of-honey-bee-colonies/. "Two widely used neonicotinoids . . . appear to significantly harm honey bee colonies over the winter, particularly during colder winters. . . .The study replicated a 2012 finding from the same research group that found a link between low doses of imidacloprid and Colony Collapse Disorder (CCD), in which bees abandon their hives over the winter and eventually die. The new study also found that low doses of a second neonicotinoid, clothianidin, had the same negative effect.

Further, although other studies have suggested that CCD-related mortality in honey bee colonies may come from bees' reduced resistance to mites or parasites as a result of exposure to pesticides, the new study found that bees in the hives exhibiting CCD had almost identical levels of pathogen infestation as a group of control hives, most of which survived the winter. This finding suggests that the neonicotinoids are causing some other kind of biological mechanism in bees that in turn leads to CCD."

66 Chensheng Lu, Kenneth M. Warchol, and Richard A. Callahan, "Sub-lethal Exposure to Neonicotinoids Impaired Honey Bees Winterization before Proceeding to Colony Collapse Disorder," *Bulletin of Insectology* 67, no. 1 (2014): 67, http://www.bulletinofinsectology.org/pdfarticles/vol67-2014-125-130lu.pdf.

67 Bjorn Rorslett, "Flowers in UltraViolet Arranged by Plant Family," NaturFotoGraf.com, http://www.naturfotograf.com/index2.html. This website allows you to view both human and insect perspectives on different flowers by clicking on images.

68 Mader et al., *Attracting Native Pollinators*, 216.

69 Ibid., 26.

70 Sara J. Wilson, *Ontario's Wealth, Canada's Future: Appreciating the Value of the Greenbelt's Eco-Services* (Vancouver: David Suzuki Foundation, 2008), 26, http://www.davidsuzuki.org/publications/downloads/2008/DSF-Greenbelt-web.pdf.

71 For a good, basic explanation of bees and apple pollination, see this classroom exercise for younger children: Camilla Barry, "Check an Apple for Pollination," Classroom Science, accessed June 12, 2014, http://www.classroomscience.org/check-an-apple-for-pollination.

72 Wilson, *Ontario's Wealth*, 26. See also page 28 for a table showing the value of the greenbelt's pollination services by natural cover type.

73 For a summary of what this insecticide is, why we're concerned about DDT, its harmful effects, how we're exposed to it and where it can still be found (even though it was banned in 1972), see "DDT," United States Environmental Protection Agency, accessed June 12, 2014, http://www.epa.gov/pbt/pubs/ddt.htm.

74 Stephen L. Buchmann and Gary Paul Nabhan, *The Forgotten Pollinators* (Washington, DC: Island Press, 1996), 198–99.

75 M.F. Mitchell and J.F. Roberts, "A Case Study of the Use of Fenitrothion in New Brunswick: The Evolution of an Ordered Approach to Ecological Monitoring," in *Effects of Pollutants at the Ecosystem Level*, ed. P.J. Sheehan, D.R. Miller, G.C. Butler and P.H. Bourdeau (Toronto: John Wiley & Sons, 1984), 397. "The low-bush blueberry, dependent on bees for pollination, is now known to be adversely affected by the fenitrothion spray programme to such an extent that compensation has been paid to growers (e.g., $58,000 to Bridges Brothers Ltd.; The Supreme Court of New Brunswick, 1976)."

76 Candace Savage, "The Plight of the Bumble Bee," *Canadian Geographic*, December 2008, 57, http://www.uoguelph.ca/canpolin/Publications/Canadian%20Geographic%20December%202008.pdf.

77 Krishna Ramanujan, "Insect Pollinators Contribute $29 billion to U.S. Farm Income," *Cornell Chronicle*, May 22, 2012, http://www.news.cornell.edu/stories/2012/05/insect-pollinators-contribute-29b-us-farm-income.

78 "Economic Value of Pollinators," Environmental Justice Foundation, accessed June 14, 2014, http://www.ejfoundation.org/bees/economic_value_of_pollinators.

79 "Global Prices of Pollination-dependent Products such

as Coffee and Cocoa Could Continue to Rise in the Long Term," press release, Helmholz Centre for Environmental Research, Research for the Environment, April 27, 2012, https://www.ufz.de/index. php?en=30403.

80 Mader et al., *Attracting Native Pollinators,* 174.

81 "Bumblebees," Seeds of Diversity, accessed October 11, 2014, http://www.seeds.ca/ pollination/pollinator-profiles/ bees/bumblebee.

82 "Bumblebee Behaviour 2: Warming up the Flight Muscles," Bumblebee.org, accessed June 14, 2014, http:// www.bumblebee.org/behaviour. htm.

83 Sheena Adams, "Green Manures: Which Cover Crops to Grow," BC Living, October 25, 2013, http://www.bcliving.ca/ garden/green-manures-which-cover-crops-to-grow.

84 Mader et al., *Attracting Native Pollinators,* 101–103.

85 "Pollinators – Busy Doing What?" Hinterland Who's Who Tube, accessed June 15, 2014, http://tv.hww.ca/video/watch/1.

86 "Bumble Bee Behaviour 2," Bumblebee.org, accessed June 24, 2014, http://www. bumblebee.org/behaviour.htm.

87 "Create a Bee-friendly Garden," David Suzuki Foundation, accessed June 15, 2014, http:// www.davidsuzuki.org/what-you-can-do/food-and-our-planet/create-a-bee-friendly-garden/. "Bees and other beneficial insects — ladybugs, butterflies, and predatory wasps — all need fresh water to drink but most can't land in a conventional bird bath without crashing. 'They're like tanks with wings,' says bee master Brian Campbell. 'They need islands in the water to touch down on.'" See also, "Bumble Bee Drinking Honey Water," YouTube video, 0:42, posted by "Culebra303," January 26, 2013, http://www.youtube.com/ watch?v=4GSGNSWvQ8o. And see, "Bees Drinking Water," YouTube video, 2:17, posted by "Splodgo," January 17, 2010, http://www.youtube. com/watch?v=8vTak5AKR8c. And finally, "Honey Bees Drinking Water in Slow Motion," YouTube video, 1:21, posted by "Home of Lucy & Mr. Muggles," August 12, 2011, http://www.youtube.com/ watch?v=sH7aNZ78FE4.

88 Neil Bowdler, "Database Shows How Bees See World in UV," BBC News, December 13, 2010, http://www.bbc.co.uk/news/ science-environment-11971274. For photos comparing how bees and humans see flowers in different colours, see Michael Hanlon, "A Bees-eye View: How Insects See Flowers Very Differently to Us," *Daily Mail* Online, August 8, 2007, http://www.dailymail.co.uk/ sciencetech/article-473897/A-bees-eye-view-How-insects-flowers-differently-us.html.

89 Rich Hatfield, Sarina Jepsen, Eric Mader, Scott Hoffmann Black and Matthew Shepherd, *Conserving Bumble Bees: Guidelines for Creating and Managing Habitat for America's Declining Pollinators* (Portland,

OR: The Xerces Society, 2012), http://www.xerces.org/bumblebees/guidelines/.

90 "Where The Wildflowers Were: Agriculture, Habitat and Biodiversity, Habitat Connectivity," Environmental Justice Foundation, accessed June 15, 2014, http://www.ejfoundation.org/bees/habitat_and_biodiversity.

91 Mader et al., *Attracting Native Pollinators*, 102.

92 "Pollinators," Canadian Wildlife Federation, accessed June 15, 2014, http://cwf-fcf.org/en/about-cwf/annual/2008-09/conservation/pollinators.html.

93 For more details on artificial native bee housing, we highly recommend The Xerces Society's beautifully illustrated guide to protecting bees and butterflies. Mader et al., *Attracting Native Pollinators*, 135–54. See also Matthew Shepherd, "Nests for Native Bees Fact Sheet," The Xerces Society for Invertebrate Conservation, accessed October 11, 2014, http://www.xerces.org/wp-content/uploads/2008/11/nests_for_native_bees_fact_sheet_xerces_society.pdf.

94 "Amazing Vandana Shiva Slams Monsanto at Rio+20 Conference 2012," YouTube video, 8:37, posted by "wingedknight," July 9, 2012, http://www.youtube.com/watch?v=Q9Zf-_LsoL8. For more on this scientist-activist, see chapter 14 in this book.

95 For the full list of both chemical and brand names, and their levels of toxicity, see Maryann Frazier, "Protecting Honey Bees from Chemical Pesticides," Pennsylvania State Beekeepers Association, accessed October 11, 2014, http://www.pastatebeekeepers.org/pdf/ProtectingBees.pdf.

96 Jennifer Hopwood, Scott Hoffman Black, Mace Vaughan and Eric Lee-Mader, *Beyond the Birds and the Bees: Effects of Neonicotinoid Insecticides on Agriculturally Important Beneficial Invertebrates* (Portland, Ore.: The Xerces Society, 2013), 4, http://www.xerces.org/wp-content/uploads/2013/09/XercesSociety_CBCneonics_sep2013.pdf.

97 Ibid., 2.

98 See J.P. Van der Sluijs et al., "Conclusions of the Worldwide Integrated Assessment on the risks of neonicotinoids and fipronil to biodiversity and ecosystem functioning," Ligue pour la Protection des Oiseaux (LPO), June 24, 2014, https://www.lpo.fr/images/Presse/cp/2014/impact_pesticides/WIA_The_following_is_a_summary_of_the_conclusions_chapter_that_will_appear_in_Environmental_Sciences_and_Pollution_Research.pdf.

99 "Harm," The Task Force on Systemic Pesticides, accessed October 12, 2014, http://www.tfsp.info/findings/harm/.

100 "Genetically Modified Soybeans: Roundup Ready Soybeans, What They Are and How They Work," SteveOntario15, accessed August 15, 2014, http://steveontario_15.tripod.com/geneticallymodifiedsoybeans/id1.html.

101   Maggie Delano, "About Roundup Ready Crops," MIT, Spring 2009, http://web.mit.edu/demoscience/Monsanto/about.html.

102   Maggie Delano, "Impact of Roundup Ready Crops," MIT, Spring 2009, http://web.mit.edu/demoscience/Monsanto/impact.html.

103   "Greenpeace Briefing," Greenpeace, accessed August 20, 2014, http://www.greenpeace.org/canada/Global/canada/report/2012/07/Oilseed-rape-impossible-to-control.pdf. See especially page 3: "Large and widespread GE oilseed rape populations have been established in the USA (Shafer et al., 2011). Two escaped GE types were found – one resistant to Monsanto's Roundup herbicide (glyphosate), and one resistant to Bayer Crop Science's Liberty herbicide (glufosinate). They also found some plants that were resistant to both herbicides, because the two GE varieties had interbred to produce a new variety with double herbicide resistance – something that was not cultivated, but formed in the wild. This echoes earlier findings in Canada where feral stacked GE traits were found (Hall et al., 2000; Orson, 2002). These GE feral oilseed rape plants could create problem weeds because of their multiple herbicide resistance. These GE populations have established since GE oilseed rape was first grown in the mid-1990s."

104   "Seeds: Monsanto Company v. Percy Schmeiser," History Commons, Genetic Engineering and the Privatization of Seeds, accessed August 20, 2014, http://www.historycommons.org/timeline.jsp?timeline=seeds_tmln&seeds_legal_actions=seeds_legalMonsantoVSchmeiser.

105   Rick Weiss, "'Gene Police' Raise Farmers Fears," *Washington Post*, February 3, 1999, http://www.mindfully.org/GE/Percy-Schmeiser-Gene-Police.htm.

106   Monsanto Canada, Inc., "Spraying Off-label Reduces Yield and Costs Farmers Money," Monsanto, April 8, 2013, http://www.monsanto.ca/newsviews/Pages/NR-2013-04-08.aspx.

107   Mark Winston, "Our Bees, Ourselves: Bees and Colony Collapse," *The New York Times*, July 14, 2014, http://www.nytimes.com/2014/07/15/opinion/bees-and-colony-collapse.html?hp&action=click&pgtype=Homepage&module=c-column-top-span-region&region=c-column-top-span-region&WT.nav=c-column-top-span-region&_r=1.

108   Hopwood et al., *Beyond the Birds*, 10.

109   Renée Johnson, *Honey Bee Colony Collapse Disorder* (Washington, DC: Congressional Research Service, January 7, 2010), 5, http://fas.org/sgp/crs/misc/RL33938.pdf.

110   *Queen of the Sun*, directed by Taggart Siegel (Portland, OR: Collective Eye, 2010), DVD.

111 Jeffri C. Bohlscheid and Frank J. Dinan, "The Case of the Missing Bees, Corn Syrup and Colony Collapse Disorder," The National Centre for Case Study Teaching in Science, accessed July 14, 2014, 2, 3 and 4, http://sciencecases.lib.buffalo.edu/cs/files/missing_bees.pdf. While moving hundreds or thousands of bee colonies as many as thousands of miles in trucks stacked four levels deep, the bees are fed high-fructose corn syrup, which contributes to their deaths. "The typical commercial bee colony is reported to travel an average of 5,500 miles per year."

112 "Pesticides + High Fructose Corn Syrup = Bee Deaths + Colony Collapse Disorder?" The Bovine (blog), May 3, 2013, http://thebovine.wordpress.com/2013/05/03/pesticides-high-fructose-corn-syrup-bee-deaths-colony-collapse-disorder/.

113 Robert Arnason, "Ontario Field Study Finds no Link between Seed Treatments, Bee Deaths," The Western Producer, February 21, 2013 http://www.producer.com/daily/ontario-field-study-finds-no-link-between-seed-treatments-bee-deaths/.

114 This item also refers to Richard J. Gill and Nigel E. Raine, "Chronic Impairment of Bumblebee Natural Foraging Behaviour Induced by Sublethal Pesticide Exposure," Functional Ecology, July 7, 2014, DOI: 10.1111/1365-2435.12292.

115 Jackson Proskow and Gil Shochat, "Ontario to Restrict Use of Neonicotinoids," Global News, July 9, 2014, http://globalnews.ca/news/903394/the-plight-of-the-bees/.

116 Russ McSpadden, "Robotic Bees to Pollinate Monsanto's Crops," Earth First, April 8, 2013, http://earthfirstjournal.org/newswire/2013/04/08/robotic-bees-to-pollinate-monsanto-crops/.

117 Mark Winston, "Our Bees, Ourselves."

118 Mader et al., Attracting Native Pollinators, 45.

Three: Taste the Connections

119 Tracy Winegar, Good Ground (Los Angeles: Omnific Publishing, 2013).

120 "Action Committee," Sierra Youth Coalition, accessed August 15, 2014, http://www.syc-cjs.org/ontario/action.

121 Darren Stott and Erin Nichols, "Local Sustainable, Food within BC Public Institutions," Public Health Association of BC, April 8, 2013, 26, http://www.phabc.org/userfiles/file/Localfoodrecommendationsi nBCPublicInstitutions-final.pdf.

122 "Backgrounder: History of VIHA-contracted Food Services," Vancouver Island Health Authority, June 23, 2005, http://www.viha.ca/NR/rdonlyres/18CDE745-B3BF-4313-8573-6B4A7BFCA2B4/0/food_svs_june05_bg.pdf.

123 Joe Murphy, "Strategic Outsourcing by a Regional

Health Authority: The Experience of the Vancouver Island Health Authority" *Healthcare Papers* 8 (September 2007): 104–111. DOI: 10.12927/hcpap.2007.19224.

124 "Compass Group Canada, Who We Are," Compass Group Canada, 2013, http://www.compass-canada.com/Pages/Content.aspx?ItemID=2.

125 Murphy, "Strategic Outsourcing."

126 For more information on David's work with this program, see Nathalie North, "Health and Healing in the Garden," *Saanich News*, October 2, 2012, http://www.saanichnews.com/news/172329941.html.

127 Sarah DeWeerdt, "Is Local Food Better?" Worldwatch Institute, August 8, 2014, http://www.worldwatch.org/node/6064.

128 Robert Arnason, "David Suzuki Takes Swipe at Conventional Agriculture," *The Western Producer*, December 5, 2012, http://www.producer.com/daily/david-suzuki-takes-swipe-at-conventional-agriculture/.

129 "Getting Wild Nutrition from Modern Food: Benefits of Grass-Fed Products," Eat Wild, accessed December 12, 2014, http://www.eatwild.com/healthbenefits.htm.

130 John and Ocean Robbins, "The Truth about Grassfed Beef," The Food Revolution Network, December 19, 2012, http://foodrevolution.org/blog/the-truth-about-grassfed-beef/. See also, H. Roger Segelken, "Simple Change in Cattle Diets Could Cut E. coli Infection," *Cornell Chronicle*, September 8, 1998, http://www.news.cornell.edu/stories/1998/09/simple-change-cattle-diets-could-cut-e-coli-infection.

131 Grace Communications Foundation, "Agriculture, Energy and Climate Change," Sustainable Table, accessed August 10, 2014, http://www.sustainabletable.org/982/agriculture-energy-climate-change.

132 The Biophysics Group at Silsoe Research Institute in England performed these tests in 2006. See "UPC Letter for Chickens in *New Haven Register*," United Poultry Concerns, November 13, 2009, http://www.upc-online.org/thinking/111306newhaven.html; and Theodosia Burr, "Chickens Know More than We Think," *Nature's Corner*, accessed June 16, 2014, http://www.naturescornermagazine.com/future_shock.html; and "Bird Brains (A Quick Overview of Chicken Intelligence)," The Southern Vegetarian, February 15, 2012, https://thesouthernvegetarian.wordpress.com/tag/chicken-intelligence/.

133 Matt McIntosh, "Ontario's Livestock Industry No Fan of 'Ag-gag' Laws," *Better Farming*, October 29, 2013, http://www.betterfarming.com/online-news/ontario%E2%80%99s-livestock-industry-no-

fan-%E2%80%98ag-gag%E2%80%99-laws-53951.

134 "For a Healthier Country, Overhaul Farm Subsidies," *Scientific American*, May 1, 2012, http://www.scientificamerican.com/article/fresh-fruit-hold-the-insulin/.

135 "Food Insecurity," *Proof*, accessed December 14, 2014, http://nutritionalsciences.lamp.utoronto.ca/food-insecurity/. See also S.I. Kirkpatrick and V. Tarasuk, "Food Insecurity and Participation in Community Food Programs among Low-income Toronto Families," *Canadian Journal of Public Health* 100, no. 2 (March/April 2009): 135–39.

136 *Hungercount 2012*, "Table 1: Food Bank Use in Canada, by Province," Food Banks Canada, accessed August 20, 2014, 4, http://foodbankscanada.ca/getmedia/3b946e67-fbe2-490e-90dc-4a313dfb97e5/hungercount2012.pdf.aspx.

137 Katie Hyslop, "BC Ties Manitoba for Highest Child Poverty," *The Tyee*, July 5, 2013, http://thetyee.ca/Blogs/TheHook/2013/07/05/BC-ties-Manitoba-for-highest-child-poverty/.

138 *2012 Child Poverty Report Card*, First Call: BC Child and Youth Advocacy Coalition, November 2012, http://www.firstcallbc.org/pdfs/EconomicEquality/First%20Call%20BC%20Child%20Poverty%20Report%20Card%202012.pdf.

139 Ibid.

140 Laura Payton, "UN Official Sparks Debate over Canadian Food Security," *CBC News*, May 16, 2012, http://www.cbc.ca/news/politics/story/2012/05/16/pol-un-canada-food-security.html.

141 Ibid.

142 Ibid.

143 "UN Food Envoy Slaps Ottawa on Scrapping Census and EU Trade Talks," *CBC News*, March 3, 2013, http://www.cbc.ca/news/politics/story/2013/03/03/pol-cp-un-food-envoy-slaps-ottawa-on-scrapping-census-and-eu-trade-talks.html.

144 Dene Moore, "BC Proposes Big Changes to Land Reserve," The Canadian Press, *Global News*, March 27, 2014, http://globalnews.ca/news/1234710/b-c-proposes-big-changes-to-land-reserve/. See also, Keith Baldrey, "Is the Sky Falling on B.C.'s Agricultural Land?" *Global News*, April 3, 2014, http://globalnews.ca/news/1249080/is-the-sky-falling-on-b-c-s-agricultural-land/. And, Yuliya Talmazan, "B.C. Farmers Rally against Changes to Agricultural Land Reserve in Victoria," *Global News*, April 7, 2014, http://globalnews.ca/news/1255792/b-c-farmers-rally-against-changes-to-agricultural-land-reserve-in-victoria/.

145 Sara J. Wilson, *Ontario's Wealth, Canada's Future: Appreciating the Value of the Greenbelt's Eco-Services* (Vancouver: David Suzuki

Foundation, 2008), 19, http://
www.davidsuzuki.org/
publications/downloads/2008/
DSF-Greenbelt-web.pdf.

146 Frank O'Brien, "Raw Land is
Where the Profit and Potential
is Being Seen Today across
Western Canada," *Western
Investor*, April 2013, http://www.
westerninvestor.com/index.php/
news/55-features/1211-raw-deals.

147 Jeff Rubin, "How Farmland
Became Canada's Hottest Real
Estate Market," *The Globe
and Mail*, September 18, 2013,
http://www.theglobeandmail.
com/report-on-business/how-
farmland-became-canadas-
hottest-real-estate-market/
article14394176/.

148 O'Brien, "Raw Land."

149 Payton, "UN Official."

150 Thomas Bosque, "7 Bottled
Water Myths – Busted," *Ban
the Bottle*, March 1, 2012,
http://www.banthebottle.net/
articles/7-bottled-water-myths-
busted.

151 Ibid. "Multinational
corporations are stepping in
to purchase groundwater and
distribution rights wherever
they can, and the bottled
water industry is an important
component in their drive to
commoditize what many feel
is a basic human right: the
access to safe and affordable
water. In the documentary
film *Thirst*, authors Alan
Snitow and Deborah Kaufman
demonstrated the rapid
worldwide privatization of
municipal water supplies, and
the effect these purchases are
having on local economies."

Four: Embrace Ecosystems

152 John Muir, with photographs
by Scot Miller, *My First
Summer in the Sierra,
Illustrated Edition* (New York:
Houghton Mifflin Harcourt,
2011), 104.

153 Luna Leopold, ed., *Round
River: From the Journals of
Aldo Leopold* (Oxford: Oxford
University Press, 1993), 145–46.

154 For teaching tools (handouts,
books and films) about Aldo
Leopold and his philosophy,
see "Fostering the Land Ethic:
Using Leopold in Teaching,"
Aldo Leopold Foundation,
accessed June 15, 2014, http://
www.aldoleopold.org/
AldoLeopold/teachingtools.
shtml. For all information
on the Aldo Leopold
Foundation, see "Aldo Leopold
Foundation," Aldo Leopold
Foundation, accessed June 15,
2014, http://www.aldoleopold.
org/home.shtml.

155 "Fast Food Statistics," Statistic
Brain, January 1, 2014, http://
www.statisticbrain.com/
fast-food-statistics/.

156 Brad Plummer, "How the U.S.
manages to waste $165 billion
in food each year," *Washington
Post*, August 22, 2012, http://
www.washingtonpost.
com/blogs/wonkblog/
wp/2012/08/22/how-food-
actually-gets-wasted-in-
the-united-states/. See also
Dana Gunders, "Wasted:
How America Is Losing Up to
40 Percent of Its Food from
Farm to Fork to Landfill," The
Natural Resources Defense
Council, August 2012, Issue
Paper (IP:12-06-B),

http://www.nrdc.org/food/files/
wasted-food-IP.pdf.

157 Eric A. Finkelstein, Justin
G. Trogdon, Joel W. Cohen
and William Dietz, "Annual
Medical Spending Attributable
to Obesity: Payer-and-Service-
Specific Estimates," *Health
Affairs*, 28, no. 5 (September/
October 2009): w822-w831,
DOI: 10.1377/hlthaff.28.5.w822.

158 "American Food Consumption
Statistics," One Green Planet,
http://www.onegreenplanet.
org/news/american-food-
consumption-statistics-
infographic/.

159 Lisa Wright, "Canadians
and Americans Spend
much Differently," *The
Star*, December 18, 2013,
http://www.thestar.com/
business/2013/12/18/canadians_
and_americans_spend_
much_differently_report.
html. See also "Comparing
and Contrasting Canadian
and American Consumers,
Some Stylized Facts," Canadian
National Household Spending
Survey, Statistics Canada, U.S.
BLS, Consumer Expenditure
Survey, December 18, 2013,
http://www.nrdc.org/food/files/
wasted-food-IP.pdf.

160 John Ikerd, "High Cost of Bad
Food or Economics of Food
Insecurity" (paper presented
at the National Heirloom
Exposition World's Pure Food
Fair, Santa Rosa, California,
September 11, 2012), http://
web.missouri.edu/ikerdj/
papers/Santa%20Rosa%20
Heirloom%20Expo%20--%20
Cost%20of%20Bad%20Food.
htm.

161 Food Addiction Research
Education (FARE) is an
excellent site that raises
"public awareness about the
physiological, genetic and
environmental factors behind
food addiction," accessed
August 22, 2014, http://
foodaddictionresearch.org/.

162 Kelly D. Brownell and Kenneth
E. Warner, "The Perils of
Ignoring History: Big Tobacco
Played Dirty and Millions
Died. How Similar Is Big
Food?" *The Milbank Quarterly*
87, no. 1 (2009): 259–94, http://
www.foodaddictionsummit.
org/documents/BrownellWa
rnerFoodTobaccoMilbank09.
pdf.

163 Barbara Kingsolver, with
Steven L. Hopp and Camille
Kingsolver, *Animal, Vegetable,
Miracle: A Year of Food Life*
(New York: HarperCollins,
2007), 115.

164 Tanya Denckla Cobb
*Reclaiming Our Food: How the
Grassroots Food Movement
Is Changing the Way We Eat*
(North Adams, Mass.: Storey
Publishing, 2011), Foreword.

165 Rachel Carson, *Silent Spring*
(New York: Houghton
Mifflin, 1962). Carson's
policy-changing book details
the effects of insecticides on
songbirds in the US – she
found that the chemicals were
killing not only insects but
birds as well: ". . . spraying
destroys not only the insects
but also their principal enemy,
the birds. When later there
is a resurgence of the insect
population, as almost always
happens, the birds are not

there to keep their numbers in check."

166 "DDT Fact Sheet," National Pesticide Information Centre, 1999, http://npic. orst.edu/factsheets/ddtgen. pdf. "DDT can still legally be manufactured in the U.S., but it can only be sold to, or used by, foreign countries." In an update since this 1999 fact sheet, Wiki (http:// en.wikipedia.org/wiki/ DDT) states: "In 2009, 3314 tonnes were produced for the control of malaria and visceral leishmaniasis, hence it still qualifies as a High Production Volume Chemical. India is the only country still manufacturing DDT, with China having ceased production in 2007. India is the largest consumer." See also Henk Van den Berg, "Global Status of DDT and Its Alternatives for Use in Vector Control to Prevent Disease" (presented at the Stockholm Convention/ United Nations Environment Programme, October 23, 2008), http://chm.pops. int/Implementation/DDT/ Meetings/DDTEG42012/ tabid/2942/mctl/ViewDetails/ EventModID/874/EventID/338/ xmid/9462/Default.aspx.

167 David Suzuki, "Biotechnology: A Geneticist's Personal Perspective," David Suzuki, pages 10–11, accessed June 16, 2014, http://www.davidsuzuki. org/david/downloads/David_ Suzuki_Biotech_essay.pdf.

168 "Biomagnification: How DDT Becomes Concentrated as it Passes Through a Food Chain," RCN Boston, August 8, 2003, http://users.rcn.com/jkimball. ma.ultranet/BiologyPages/D/ DDTandTrophicLevels.html.

169 GM Watch, "Rachel Carson Centennial: The Continuing Campaign by Monsanto & The Agri-Toxics Lobby to Silence Dissent," Organic Consumers Association, May 27, 2007, http://www.organicconsumers. org/articles/article_5402.cfm.

170 AskNature is an inspiring and extensive source of bio-friendly solutions. As the website's writers explain: "AskNature is the world's most comprehensive catalog of nature's solutions to human design challenges ... Nearly every design challenge humans face shares commonalities with the challenges the rest of life has adapted to over 3.8 billion years of 'research and development.' By understanding how these adaptations work, Innovators can mimic ideas that have thrived in balance with Earth's complex systems." See http://www.asknature. org/#menuPopup. Watch exciting, short videos of nature's solutions at http:// www.asknature.org/article/ view/nuggets. And you can sign up for AskNature's free, educational newsletter (and see old issues) at http:// biomimicry.net/connecting/ subscribe-to-our-newsletter/.

171 Liz Langley, "Animal Pharm: What Can We Learn from Nature's Self-Medicators?" *National Geographic*:

Weird and Wild, October 17, 2013, http://newswatch.nationalgeographic.com/2013/10/07/animal-pharm-what-can-we-learn-from-natures-self-medicators/.

172 Ibid.

173 Beverly Clark, "Study Finds Monarch Butterflies Use Medicinal Plants to Treat Offspring for Disease," EurekAlert press release, October 11, 2010, http://www.eurekalert.org/pub_releases/2010-10/eu-sfm100810.php.

174 Candice Hawkinson, "The Pollinators: Butterflies," Galveston County's Master Gardeners, *Beneficial in the Garden* series, 2005, http://aggie-horticulture.tamu.edu/galveston./beneficials/beneficial-66_pollinators-butterflies.htm.

175 "How Wolves Change Rivers," YouTube video, 4:34, posted by "Sustainable Man," February 13, 2014, http://sustainablehuman.com/how-wolves-change-rivers/

176 George Monbiot, "How We Ended Up Paying Farmers to Flood Our Homes," *The Guardian,* February 18, 2014, http://www.theguardian.com/commentisfree/2014/feb/17/farmers-uk-flood-maize-soil-protection.

177 George Monbiot, *Feral: Searching for Enchantment on the Frontiers of Rewilding* (London: Allen Lane, 2013). For more on George Monbiot, visit http://www.monbiot.com/.

178 See http://sustainablehuman.com/

179 Jacob Chamberlain, "Farmers Rise Up Against Agribusiness, Face Down Riot Police in Brazil," *Common Dreams,* February 13, 2014, http://www.commondreams.org/headline/2014/02/13-4.

180 See *Resurgence & Ecologist* magazine online at http://www.resurgence.org/magazine/.

181 Satish Kumar, "Learning from Nature," *Resurgence & Ecologist*, accessed June 15, 2014, https://www.resurgence.org/satish-kumar/articles/learning-from-nature.html.

182 "Postgraduate Courses," Schumacher College, accessed June 15, 2014, http://www.schumachercollege.org.uk/courses/postgraduate-courses.

183 George Sylvester Viereck, interview, "What Life Means to Einstein," *Saturday Evening Post*, October 26, 1929, accessed June 15, 2014, http://www.saturdayeveningpost.com/wp-content/uploads/satevepost/what_life_means_to_einstein.pdf. See also Jeff Nilsson, "'Imagination Is More Important than Knowledge,'" *Saturday Evening Post*, Post Perspective, accessed June 15, 2014, http://www.saturdayeveningpost.com/2010/03/20/archives/post-perspective/imagination-important-knowledge.html.

184 For more on the benefits of group flight by geese (as well as pelicans and other heavier birds), see Dr. Robert McNeish,

"Lessons from the Geese," ARPA Adventure Farm, accessed June 15, 2014, http://arpa.co.za/wildgeese.html. For more on the dynamics of air flow, see "The Work of Wings," National Aeronautics and Space Administration, Virtual Skies, accessed June 15, 2014, http://virtualskies.arc.nasa.gov/aeronautics/3.html.

185 Robin McClary, "About Canada Geese: Nature, Description, and Behavior of Canada Geese," Citizens for the Preservation of Wildlife, 2004, http://www.preservewildlife.com/geeseworld.htm.

186 "7 Lessons We Can Learn from Geese to Succeed at Work," 7 Geese, July 12, 2012, http://blog.7geese.com/2012/07/12/7-lessons-we-can-learn-from-geese-to-succeed-at-work/. See also "Lessons of the Geese," YouTube video, 3:24, posted by "In Search of Me Cafe," September 2, 2010, http://www.youtube.com/watch?v=hazitrxzhPk.

187 Joan Y. Edwards, "What You Can Learn from a Deer," Joan Y. Edwards, May 30, 2012, http://joanyedwards.wordpress.com/2012/05/30/what-you-can-learn-from-a-deer/.

188 "White-tailed Deer," Fairfax Country Public Schools, accessed June 16, 2014, http://www.fcps.edu/islandcreekes/ecology/white-tailed_deer.htm.

189 "White-tailed Deer," National Geographic, accessed June 16, 2014, http://animals.nationalgeographic.com/animals/mammals/white-tailed-deer/.

190 "Living with Wildlife: Deer," Washington Department of Fish and Wildlife, accessed June 16, 2014, http://wdfw.wa.gov/living/deer.html.

191 "Sitka Blacktail Deer," Super Slam of North American Big Game, accessed June 16, 2014, http://www.superslam.org/know-your-game/sitka-blacktail-deer.

192 "The Private Life of Deer," PBS, infographic, accessed June 16, 2014, http://www.pbs.org/wnet/nature/episodes/the-private-life-of-deer/infographic-learn-about-the-whitetailed-deer/8314/.

193 "Living with Wildlife: Deer."

194 "What's the Use? The Reasons for Preserving Biodiversity Are Becoming More Widely Understood," The Economist, September 14, 2013, http://www.economist.com/news/special-report/21585093-reasons-preserving-biodiversity-are-becoming-more-widely-understood-whats-use.

195 "Frequently Asked Questions about Bylaw Enforcement, Bylaw 8200," District of Saanich, accessed June 16, 2014, http://www.saanich.ca/living/pdf/bylawenforcementbrochure.pdf.

196 See Canadian Gardening's article on top-ten climbing roses for more information. Judith Adam, "Top Ten Climbing Roses," Canadian Gardening, accessed March 13, 2015, http://

www.canadiangardening.
com/plants/roses/
top-10-climbing-roses/a/1452.

197 Adapted from an interview
with Nathalie Chambers by
Judy Lavoie, "Farmer Counts
Deer, Geese as Friends," *Times
Colonist*, December 16, 2012,
http://www.deerfriendly.
com/international-news/
canada-deer/herd-
population-and-management/
farmer-counts-deer-geese-
as-friends-december-16-2012-
british-columbia.

198 Wade Davis, "The World
Until Yesterday by Jared
Diamond – review," *The
Guardian*, January 9, 2013,
http://www.guardian.co.uk/
books/2013/jan/09/history-
society. See also Wade Davis,
*The Wayfinders: Why Ancient
Wisdom Matters in the Modern
World* (Toronto: House of
Anansi Press, 2009).

199 Wade Davis, *The Wayfinders:
Why Ancient Wisdom Matters
in the Modern World*
(Toronto: House of Anansi
Press, 2011).

200 Bianca Jagger, "Women, the
Unsung Heroes," *Huffington
Post*, March 8, 2013, http://
www.huffingtonpost.com/
bianca-jagger/women-the-
unsung-heroes-0_b_2838414.
html.

201 Colin Stief, "Slash and Burn
Agriculture: Slash and
Burn Can Contribute to
Environmental Problems,"
About.com Geography,
accessed June 16, 2014, http://
geography.about.com/od/
urbaneconomicgeography/a/
slashburn.htm.

202 Dr. Gerald Urquhart, "The
Virtual Rainforest," The
Virtual Rainforest, Michigan
State University, accessed June
16, 2014, https://www.msu.
edu/user/urquhart/rainforest/
Content/Neotropical-
Rainforest.html. "The tropical
rainforest [is] the most diverse
ecosystem on Earth, with an
estimated 70% of the Earth's
species."

203 "Scientists Estimate the Age of
Trees in the Amazon," Public
Affairs and Communications
University of California, Santa
Barbara, news release, January
7, 1998, http://www.ia.ucsb.
edu/pa/display.aspx?pkey=161.
"In a breeding population of
trees with ages ranging from
decades to millennia, unique
genes that might be lost with
the passing of generations are
maintained in the long-lived
individuals ... Thousand-
year-old trees can reproduce
with trees that are many
generations younger and
prevent the loss of genetic
diversity."

204 Stief, "Slash and Burn."

205 K. Kris Hirst, "Slash and Burn:
Agricultural System Called
Slash and Burn," About.com
Archaeology, accessed June
16, 2014, http://archaeology.
about.com/od/skthroughsp/qt/
slash_burn.htm.

206 "Slash-and-Burn Agriculture,"
Encyclopaedia Britannica,
accessed June 16, 2014,
http://www.britannica.com/
EBchecked/topic/548086/
slash-and-burn-agriculture.

207 John Muir, *Our National
Parks*, "Chapter 10: The

American Forests," Sierra Club, accessed June 16, 2014, http://vault.sierraclub.org/john_muir_exhibit/writings/our_national_parks/.

208 Nancy J. Turner, *The Earth's Blanket: Traditional Teachings for Sustainable Living* (Vancouver: Douglas & McIntyre, 2005), 14.

Five: Encourage Biodiversity

209 Frans Lanting in conversation with Diedtra Hendersen, "A Natural Eye: Wildlife Photographer Frans Lanting Tries to Adapt to the World of His Subjects," *The Seattle Times*, February 22, 1999, http://community.seattletimes.nwsource.com/archive/?date=19990222&slug=2945591.

210 Rick Cheeseman, "A Response to the Discussion Paper by the Agricultural Land Review Committee," Pinnacle Farms, February 24, 2010, http://pinnaclefarms.ca/ALRCinput.pdf. Cheeseman prepared this slideshow in response to Nova Scotia, Canada's Provincial Interest Statements on Agricultural Land, whose goals were defined as "protection of agricultural land" and "viable and sustainable food resource base."

211 Bianca Jagger, "Women, the Unsung Heroes," *Huffington Post*, March 8, 2013, http://www.huffingtonpost.com/bianca-jagger/women-the-unsung-heroes-o_b_2838414.html. "Traditional indigenous cultures use natural resources sustainably . . . 'native peoples have been stewards of 99 percent of the world's genetic resources.' Women of the forest-dwelling Kpelle tribe of Liberia, for instance, sow more than 100 varieties of rice, making their fields a wealth of genetic diversity."

212 For information about Betty Krawczyk, see http://en.wikipedia.org/wiki/Betty_Krawczyk, and http://bettysearlyedition.blogspot.ca/.

213 See http://www.womeninthewoods.com/.

214 For news and videos with current perspectives on the original Clayoquot Sound protest, see David Tindall, "Twenty Years after the Protest: What We Learned from Clayoquot Sound," *The Globe and Mail*, August 12, 2013, http://www.theglobeandmail.com/commentary/twenty-years-after-the-protest-what-we-learned-from-clayoquot-sound/article13709014/?page=all. See also Dionne Bunsha, "What Clayoquot Sound Faces Now," *The Tyee*, August 19, 2013, http://thetyee.ca/Opinion/2013/08/19/Clayoquot-Faces-Now/; and Kim Nurdall, "Twenty Years Later: The 'War in the Woods' at Clayoquot Sound Still Reverberates across B.C.," *Global BC*, August 11, 2013, http://globalnews.ca/news/774070/twenty-years-later-the-war-in-the-woods-at-clayoquot-sound-still-

reverberates-across-b-c/. For the Friends of Clayoquot Sound website, go to http://focs.ca/.

215 For information on the desecration of this amazingly beautiful, natural, spiritual location, called *Spaet* by the First Nations (pronounced "Spa-eth," meaning "bear"), see "Spaet," First Nations: Land Rights and Environmentalism in British Columbia, accessed June 17. 2014, http://www.firstnations. de/development/coast_salish-spaet.htm.

216 For an overview of Spencer's Pond wildlife habitat photos, videos and links regarding the impact of the Spencer Interchange, see http://www.spencerspond.ca/.

217 Cheeseman, "A Response," 17.

218 Ibid., 10.

219 Ibid., 11.

220 Ibid.

221 Ibid., 12.

222 Robert Arnason, "David Suzuki Takes Swipe at Conventional Agriculture," *The Western Producer*, December 5, 2012, http://www.producer.com/daily/david-suzuki-takes-swipe-at-conventional-agriculture/.

223 Terrence Veeman and Michele Veeman, "Agriculture and Food," *The Canadian Encyclopedia*, accessed June 17. 2014, http://www.thecanadianencyclopedia.ca/en/article/agriculture-and-food/.

224 National Farmers Union, "The Farm Crisis: Its Causes and Solutions," National Farmer's Union, July 5, 2005, http://www.nfu.ca/sites/www.nfu.ca/files/Ministers_of_Ag_brief_FOUR.pdf. See especially page 8. See also "Putting the Cartel before the Horse ... and Farm, Seeds, Soil, Peasants, etc.: Who Will Control Agricultural Inputs, 2013?" Etc Group, September 2013, http://www.etcgroup.org/sites/www.etcgroup.org/files/CartelBeforeHorse11Sep2013.pdf.

225 "Putting the Cartel before the Horse." "The six companies are Monsanto, DuPont, Syngenta, Bayer, Dow and BASF. Note that BASF is not included among the top 10 seed companies. While the company does not have significant retail seed sales, it is heavily engaged in seed research and has partnerships with several of the other five companies and investments in several start-up enterprises."

226 "Tunis 2013: If We Rely on Corporate Seed, We Lose Food Sovereignty," Etc Group, April 4, 2013, http://www.etcgroup.org/content/tunis-2013-if-we-rely-corporate-seed-we-lose-food-sovereignty.

227 Ibid.

228 La Via Campesina (http://viacampesina.org/en/) is an international social justice movement formed in 1993, composed of millions of peasants, farmers, Indigenous peoples, migrants, agricultural workers, landless people and women farmers from 73

countries who got together to accomplish food sovereignty in their communities and to promote sustainable, healthy food for the world. Guy Kastler helped create the Peasant Seeds Network, which advocates self-seed agricultural practices.

229 "Tunis 2013."

230 "Farm Facts or Fiction," Pinnacle Farms, accessed June 17, 2014, http://www.pinnaclefarms.ca/FarmFactsOrFiction/FarmFactsOrFiction.html.

231 John Muir, with photographs by Scot Miller, *My First Summer in the Sierra, Illustrated Edition* (New York: Houghton Mifflin Harcourt, 2011), 14, 16.

232 Derek Thompson, "Cheap Eats: How America Spends Money on Food," *The Atlantic*, March 8, 2013, http://www.theatlantic.com/business/archive/2013/03/cheap-eats-how-america-spends-money-on-food/273811/. See also, "Food Source: Farm Size and Ownership," U.S. Farmers and Ranchers Alliance, The Food Dialogues, accessed June 18, 2014, http://www.fooddialogues.com/foodsource/farm-size-and-ownership. Similar figures are also documented by James M. MacDonald, "Why Are Farms Getting Larger? The Case of the U.S." (address, annual meeting of the German Association of Agricultural Economists [GeWiSoLa], Halle, Germany, September 28, 2011), 39, http://ageconsearch. umn.edu/bitstream/115361/2/MacDonald.pdf.

233 Thompson, "Cheap Eats."

234 "Food Source: Farm Size and Ownership."

235 "Fun Facts about the Food We Eat," National Ag Day, accessed June 17, 2014, http://www.agday.org/education/fun_facts.php.

236 MacDonald, "Why Are Farms Getting Larger?" 3.

237 "Factory Farm Nation: Map Charts Unprecedented Growth in Factory Farming," Food & Water Watch, November 30, 2010, http://www.foodandwaterwatch.org/pressreleases/factory-farm-nation-map-charts-unprecedented-growth-in-factory-farming/.

238 "Factory Farming in Canada," BeVeg.ca, accessed October 11, 2014, http://www.beveg.ca/factory-farming-in-canada.php.

239 Rachel Cernansky, "Factory Farms Decreasing in Number But Increasing in Size: 20% Growth in 5 Years," *Treehugger*, November 30, 2010, http://www.treehugger.com/corporate-responsibility/factory-farms-decreasing-in-number-but-increasing-in-size-20-growth-in-5-years.html.

240 "Agriculture in the United States," Wikipedia, accessed June 17, 2014, http://en.wikipedia.org/wiki/Agriculture_in_the_United_States. See also Bureau of Labor Statistics, "Employment by Major Industry Sector,"

United States Department of Labor, December 2013, http://www.bls.gov/emp/ep_table_201.htm; and National Institute of Food and Agriculture, "About Us," United States Department of Agriculture, accessed June 17, 2014, http://www.csrees.usda.gov/qlinks/extension.html.

241 Grace Communications Foundation, "Waste Management: Mountains of Manure," Sustainable Table, accessed October 13, 2014, http://www.sustainabletable.org/906/waste-management.

242 "United States Facts," Factory Farm Map, accessed October 12, 2014, http://www.factoryfarmmap.org/states/us/.

243 Ibid.

244 Ibid.

245 *Curb the Cruelty: Canada's Farm Animal Transport System in Need of Repair* (Toronto: World Society for the Protection of Animals, 2010), 4, http://www.worldanimalprotection.ca/ati/CurbtheCrueltyReport.pdf.

246 *What's On Your Plate? The Hidden Costs of Industrial Animal Agriculture in Canada* (Toronto: World Society for the Protection of Animals, 2012), 16, http://richarddagan.com/cafo-ilo/WSPA_WhatsonYourPlate_FullReport.pdf.

247 *Curb the Cruelty*, 6.

248 "Farm Animals," Animal Alliance of Canada, accessed August 29, 2014, http://www.animalalliance.ca/campaigns/farm-animals.html.

249 MacDonald, "Why are Farms Getting Larger?" 3.

250 Ibid., 43.

251 Ibid., 31–32.

252 "Industrial Agriculture and Small-scale Farming," Global Agriculture, accessed June 17, 2014, http://www.globalagriculture.org/report-topics/industrial-agriculture-and-small-scale-farming/industrial-agriculture-and-small-scale-farming.html.

253 Beverly D. McIntyre, Hans R. Herren, Judi Wakhungu and Robert T. Watson, eds., *Agriculture at a Crossroads: International Assessment of Agricultural Knowledge, Science and Technology for Development Synthesis Report, A Synthesis of the Global and Sub-global IAASTD Reports* (Washington, DC: Island Press, 2009), 17, accessed June 18, 2014, http://www.unep.org/dewa/agassessment/reports/IAASTD/EN/Agriculture%20at%20a%20Crossroads_Synthesis%20Report%20(English).pdf.

254 Ibid., 3. Italics added for emphasis.

255 "Dr. Lee's Interview with Harold Steves," June 7, 2014, Brio Integrative Health Centre Inc., http://www.yourbriohealth.com/news/dr-lees-interview-harold-steves

256 Veeman and Veeman, "Agriculture and Food."

257 "Why Incorporate?" INC Business Lawyers, accessed June 17, 2014, www.incorporate.ca/why-incorporate.

258 Susan Ward, "7 Reasons to Incorporate Your Business," About.com, Small Business: Canada, accessed June 17, 2014, http://sbinfocanada.about.com/od/incorporation/a/incorporate.htm.

259 "Six Powerful Reasons to Incorporate or Organize an LLC," AmeriLawyer.com, accessed June 18, 2014, https://www.amerilawyer.com/sixreasons.htm.

260 National Farmers Union, "The Farm Crisis: Its Causes and Solutions," National Farmers Union, July 5, 2005, http://www.nfu.ca/sites/www.nfu.ca/files/Ministers_of_Ag_brief_FOUR.pdf. See especially pages 3 and 9.

261 Ibid., "The Farm Crisis," 3.

262 Ibid., 5, 6, 7, 8 and 15.

263 Veeman and Veeman, "Agriculture and Food."

264 For images that pictorially define a dead zone, see "Special Report – Dead Zones: Mysteries of Ocean Die-offs Revealed," National Science Foundation, accessed October 10, 2014, http://www.nsf.gov/news/special_reports/deadzones/index.jsp.

265 Johns Hopkins Bloomberg School of Public Health and Shawn McKenzie, A Brief History of Agriculture and Food Production: The Rise of "Industrial Agriculture" (Baltimore: Johns Hopkins University, 2007), 23, https://static.squarespace.com/static/508e3b97e4b047ba54da61b9/t/51317845e4b060819b0dd8a1/1362196549046/Shawn%20McKenzie%20MPH_A%20Brief%20History%20of%20Agriculture%20and%20Food%20Production.pdf.

266 Bryan Walsh, "This Year's Gulf of Mexico Dead Zone Could Be the Biggest on Record," Time, Science and Space, June 19, 2013, http://science.time.com/2013/06/19/this-years-gulf-of-mexico-dead-zone-could-be-the-biggest-on-record/. See also Bryan Walsh, "A Smaller than Predicted Dead Zone is Still Toxic for the Gulf of Mexico," Time, Science and Space, July 30, 2013, http://science.time.com/2013/07/30/a-smaller-than-predicted-dead-zone-is-still-toxic-for-the-gulf-of-mexico/.

267 Tom Philpott, "Why This Year's Dead Gulf Zone Is Twice as Big as Last Year's," Mother Jones, August 14, 2013, http://www.motherjones.com/tom-philpott/2013/08/gulf-of-mexico-dead-zone-growth.

268 Ibid.

269 At Virginia Institute of Marine Science, one can view a graph entitled "Dead Zones by Decade from 1910 through 2007." See "Trends, Low-oxygen 'Dead Zones' are Increasing around the World," Virginia Institute of Marine Science, accessed October 10, 2014, http://www.vims.edu/research/topics/dead_zones/trends/index.php.

270 "Dead Zone (ecology)," Wikipedia, accessed October 14, 2014, http://en.wikipedia.org/wiki/Dead_zone_(ecology).

271 Steven Kennedy, "Ocean Dead Zones Infographic," DistanceLearning.com, accessed June 18, 2014, http://www.distancelearning.com/resources/oceanic-dead-zones-infographic/.

272 "Hidden Costs of Industrial Agriculture," Union of Concerned Scientists, accessed June 18, 2014, http://www.ucsusa.org/food_and_agriculture/our-failing-food-system/industrial-agriculture/hidden-costs-of-industrial.html.

273 Vandana Shiva, Debbie Barker and Caroline Lockhart, coordinators, *The GMO Emperor Has No Clothes: A Global Citizens Report on the State of GMOs – False Promises, Failed Technologies*, accessed June 18, 2014, 229, http://www.navdanya.org/attachments/Latest_Publications9.pdf. See also "Horizontal Gene Transfer from GMOs Does Happen," The Institute of Science in Society, accessed October 12, 2014, http://www.i-sis.org.uk/horizontalGeneTransfer.php.

274 Shiva, Barker and Lockhart, *GMO Emperor*, 222.

275 Ibid., 247. "The *Safe Shoppers Bible* says that Monsanto's Ortho Weed-B-Gon Lawn Weed Killer contains a known carcinogen, 2, 4 D."

276 Shawn McKenzie, MPH, *A Brief History of Agriculture and Food Production: The Rise of "Industrial Agriculture,"* Johns Hopkins Bloomberg School of Public Health, 2007, accessed June 18, 2014, 27, https://static.squarespace.com/static/508e3b97e4b047ba54da61b9/t/51317845e4b060819b0d d8a1/1362196549046/Shawn%20McKenzie%20MPH_A%20Brief%20History%20of%20Agriculture%20and%20Food%20Production.pdf.

277 Leslie Beck, "New Study Finds Scant Evidence of Health Benefits from Eating Organic Foods," *The Globe and Mail*, September 4, 2012, http://www.theglobeandmail.com/life/health-and-fitness/health/new-study-finds-scant-evidence-of-health-benefits-from-eating-organic-foods/article4517773/.

278 "New Challenges/New Choices," Freeman Spogli Institute for International Studies, http://iis-db.stanford.edu/pubs/23434/FSI_2011ARFinal_Low.pdf.

279 "Scientists Tied to Tobacco Industry Propaganda, and Funding from Monsanto, Turn Attention to Organic Food," The Cornucopia Institute, September 12, 2012, accessed June 18, 2014, http://www.cornucopia.org/2012/09/stanfords-spin-on-organics-allegedly-tainted-by-biotechnology-funding/.

280 Barry Estabrook, *Tomatoland: How Modern Industrial Agriculture Destroyed Our Most Alluring Fruit* (Kansas City: Andrews McMeel Publishing, 2012).

281 "Dead Zones Need Immediate Attention: Lack of Oxygen in Coastal Waters will Create Social, Economic and Recreational Problems If Not Addressed Faster, a New Report Says," United Nations

Environment Programme, October 2011, http://www.unep.org/dgef/Portals/43/news/stories/IWC6DeadZone.pdf.

282 The National Farmers Union, "The Farm Crisis," 12–20.

283 Ibid.

284 Ibid., 20.

285 Janine M. Benyus, *Biomimicry: Innovation Inspired by Nature* (New York: HarperCollins, 1997). See also Benyus's TED talk, "Biomimicry in Action," TED, accessed June 18, 2014, http://www.ted.com/talks/janine_benyus_biomimicry_in_action.html. Finally, see "What Do You Mean by the Term Biomimicry: A Conversation with Janine Benyus," Biomimicry Institute, accessed June 18. 2014, http://www.biomimicryinstitute.org/about-us/what-do-you-mean-by-the-term-biomimicry.html.

286 Benyus, *Biomimicry*, 3. See also a discussion regarding failures and fossils in "What *is* Biomimicry?" Biomimicry Institute, accessed June 18, 2014, http://biomimicryinstitute.org/about-us/what-is-biomimicry.html.

287 Benyus, *Biomimicry*, 5–6.

288 Valerie Solheim, *Bees Healing Bees,* Healing Bees, April 4, 2009, http://www.healingbees.org/resources/Beeshealingbees+report.pdf.

289 "Feldenkrais Method," International Feldenkrais Federation, accessed June 19, 2014, http://feldenkrais-method.org/en/feldenkrais-method.

290 Daven Hiskey, "Velcro was Modeled after the Burrs of the Burdock Plant That Stuck to Velcro Inventor's Pants After a Hunting Trip," Today I Found Out, accessed August 12, 2014, http://www.todayifoundout.com/index.php/2011/09/velcro-was-modeled-after-burrs-of-the-burdock-plant-that-stuck-to-velcros-inventors-pants-after-a-hunting-trip/.

291 "Meet Jeremy Baker," Spin the Black Disc, accessed June 19, 2014, http://spintheblackdisc.com/about/.

Six: Transcend the Paradox

292 Dr. Steve Maraboli, "Inspirational Quotes," accessed June 19, 2014, http://www.stevemaraboli.com/Inspirational-Quotes.html.

293 "2013 Conference," International National Trusts Organisation, accessed June 19, 2014, http://internationaltrusts.org/programmes/2013-conference.

294 Therese J. Borchard, "Five Emotional Vampires and How to Combat Them," Psych Central, World of Psychology, accessed June 19, 2014, http://psychcentral.com/blog/archives/2009/10/27/5-emotional-vampires-and-how-to-combat-them/. See also Judith Orloff, *Emotional Freedom: Liberate Yourself from Negative Emotions and Transform Your Life* (New York: Harmony, 2009).

295 Marvin Garbeh Davis, *Brave New Child: Liberating the Children of Liberia and the*

World, Lessons in Peace Education from a War-torn Country (Liberia: Common Ground Society; Victoria, BC: Trafford Publishing, 2009). Available through Youth Peace Literacy, http://www.atriumsoc.org/adult-brave-new-child-liberia.php.

296 See the Atrium Society (Youth Peace Literacy) website and its free, award-winning, peace-educating resources for youth, adults, parents and educators, http://www.atriumsoc.org/. The Atrium Society also provides training programs in understanding and alleviating the fundamental source of human conflict – conditioned thinking – by using observant, aware thinking in each fresh moment.

297 Janine M. Benyus, *Biomimicry* (New York: HarperCollins, 1997), 3.

298 "About Food Secure Canada," Food Secure Canada, accessed June 19, 2014, http://foodsecurecanada.org/about-us. "Food Secure Canada is a national voice for the food security movement in Canada. It is a non-profit organization with individual and organization members across Canada. The organization is based in three interlocking commitments to Zero Hunger; Healthy and Safe Food; A Sustainable Food System."

299 "About the National Farmers Union," National Farmers Union, accessed June 19, 2014, http://www.nfu.ca/about/about-national-farmers-union. "The National Farmers Union is a direct-membership organization made up of Canadian farm families who share common goals. . . . Our goal is to work together to achieve agricultural policies which will ensure dignity and security of income for farm families while enhancing the land for future generations."

300 Centre for Science in the Public Interest, accessed June 19, 2014, http://www.cspinet.org/canada/. "The Centre for Science in the Public Interest (CSPI) is an independent health advocacy organization with offices in Ottawa and Washington. . . . For more than a decade, CSPI has been an important force in stimulating changes to government policies and industry practices to improve the nutritional quality of the Canadian food supply and, most importantly, the health of Canadians."

301 "About ITK," ITK, accessed June 19, 2014, https://www.itk.ca/about-itk. "Inuit Tapiriit Kanatami (ITK) . . . is the national voice of 55,000 Inuit living in 53 communities across the Inuvialuit Settlement Region (Northwest Territories), Nunavut, Nunavik (Northern Quebec), and Nunatsiavut (Northern Labrador) land claims regions. . . . Founded in 1971 ITK represents and promotes the interests of Inuit on a wide variety of environmental, social, cultural and political, issues and challenges facing Inuit on the national level."

PART TWO: TAKE ACTION

Seven: Overcome the Obstacles

302 See http://www.jamesshuggins. com/h/quo1/quotations_ native_americans.htm.

303 See http://internationaltrusts. org/.

304 Susan Pigg, "Agricultural Land Prices Have Soared across Canada, as Demand Outstrips Supply," Simcoe. com, September 12, 20134, http://www.simcoe.com/ news-story/4075998- agricultural-land-prices-have- soared-across-canada-as- demand-outstrips-supply/.

305 Joel Schlesinger, "Rising Prices Curb Dream of Buying Farmland," *The Globe and Mail*, September 9, 2013, http://www. theglobeandmail.com/report- on-business/industry-news/ property-report/rising-prices- curb-dream-of-buying-land/ article14196607/.

306 Barrie McKenna, "Green Acres: The Soaring Value of Canada's Farmland," *The Globe and Mail*, September 6, 2013, http://www.theglobeandmail. com/report-on-business/ green-acres-the-soaring- value-of-canadas-farmland/ article14154870/.

307 Schlesinger, "Rising Prices."

308 McKenna, "Green Acres."

309 Christen Croley, "Saving Farmland: Land Conservation Aids Local Communities," *The Sentinel*, July 13, 2013, http://cumberlink.com/news/ local/saving-farmland-land- conservation-aids-local- communities/article_cfocc6ca-

eco8-11e2-beca-0019bb2963f4. html.

310 "How the ALR was Established," Agricultural Land Commission, accessed June 20, 2014, http://www.alc. gov.bc.ca/alr/Establishing_ the_ALR.htm.

311 See http://www.alc.gov.bc.ca/.

312 See http://farmlandprotection. ca/ and the group's Facebook page at https://www.facebook. com/FarmlandProtection.

313 David Friend, "Growing Young Farmers," brochure, accessed December 19, 2014, http://static.squarespace. com/static/ 547a96e6e4b09ed1ced8bc6c/ t/548df8c4e4b0a269c5995e35 /1418590404838/112514+GYF S+Brochure.pdf. "We believe that all school students should learn and be encouraged to grow health-friendly food."

314 Cindy E. Harnett, "Crowd Rallies at Legislature to Support Agricultural Land Reserve," *Times Colonist*, February 10, 2014, http:// www.timescolonist.com/ crowd-rallies-at-legislature- to-support-agricultural-land- reserve-1.845316.

315 "ALR 101," Farmland Protection Coalition, accessed June 21, 2014, http:// farmlandprotection.ca/alr-101/.

316 "Bill 24 – 2014, Agricultural Land Commission Amendment Act, 2014," Legislative Assembly of British Columbia, accessed June 21, 2014, http://www. leg.bc.ca/40th2nd/1st_read/ gov24-1.htm.

317 Mark Hume, "Leading B.C. Scientists Criticize Plans for Agriculture Land Reserve," *The Globe and Mail*, April 10, 2014, http://www.theglobeandmail.com/news/british-columbia/leading-bc-scientists-criticize-plans-for-agricultural-land-reserve/article17909527/.

318 Judith Lavoie, "ALR Change Raises Suspicions," *Focus Online*, May 2014, http://www.focusonline.ca/?q=node/720.

319 "BC Food Systems Network Statement on the Passage of Bill 24," BC Food Systems Network, accessed August 20, 2014, http://bcfsn.org/what-we-do/protecting-the-agriculture-land-reserve.

320 Charles Campbell, "Forever Farmland: Reshaping the Agricultural Land Reserve for the 21st Century" (Vancouver: David Suzuki Foundation, 2006), http://www.davidsuzuki.org/publications/downloads/2006/DSF-ALR-final3.pdf.

321 Ibid., 5.

322 See another article by Charles Campbell, "Is BC Down on the Farm? In the East Kootenays, Critics See More Proof the Agricultural Land Reserve Isn't Working," *The Tyee*, May 30, 2006, http://thetyee.ca/News/2006/05/30/DownOnTheFarm/.

323 Leslie Puska, Liz Clements and Kelsey Chandler, Office of Community Based Research, "Climate Change and Food Security on Vancouver Island: Discussion Paper," August 2011, http://www.uvic.ca/research/centres/cue/assets/docs/Climate%20Change%20and%20Food%20Report.pdf

324 See chapter 12 for more on land trusts.

325 "Norma Lohbrunner Will Be Sadly Missed. Celebration of Life, October 29, 2011," The Land Conservancy, accessed June 21, 2014, http://blog.conservancy.bc.ca/2011/10/norma-lohbrunner-will-be-sadly-missed-celebration-of-life-october-29-2011/.

326 See "Joseph Lohbrunner Farm and Bird Sanctuary," The Land Conservancy, accessed June 21, 2014, http://blog.conservancy.bc.ca/properties/vancouver-island-region/joseph-lohbrunner-farm-and-bird-sanctuary/. For lovely views of the Lohbrunner farm and farmhouse, see "Joseph Lohbrunner Bird Sanctuary and Farm, Langford, BC," YouTube video, 2:37, posted by "TKCAdmin's channel," August 14, 2012, http://www.youtube.com/watch?v=2qyaOnwaPME.

327 "Turtle Valley Farm," The Land Conservancy, accessed June 21, 2014, http://blog.conservancy.bc.ca/properties/okanagan-region/turtle-valley-farm/.

328 Stephen D. Simpson, "The Banking System: Commercial Banking – How Banks Make Money," Investopedia, accessed August 20, 2014, http://www.investopedia.com/university/banking-system/banking-system3.asp. "Deposits: The largest

source by far of funds for [US] banks is deposits; money that account holders entrust to the bank for safekeeping and use in future transactions, as well as modest amounts of interest. Generally referred to as 'core deposits,' these are typically the checking and savings accounts that so many people currently have. . . . While people will typically maintain accounts for years at a time with a particular bank, the customer reserves the right to withdraw the full amount at any time. Customers have the option to withdraw money upon demand and the balances are fully insured, up to $250,000, therefore, banks do not have to pay much for this money."

329 For an example of TLC's financial statements, available for all to see on its website, see "Consolidated Financial Statements: TLC The Land Conservancy of British Columbia," The Land Conservancy, April 30, 2011, http://blog.conservancy.bc.ca/wp-content/uploads/2011/09/TLC-11cfs_No-DRAFT-watermark.pdf.

330 To view a list of properties acquired from 1997 until 2010, see "Land Protection," The Land Conservancy, accessed June 21, 2014, http://blog.conservancy.bc.ca/wp-content/uploads/2010/12/2010-Land-Protection.pdf.

331 "TLC's Q&A from Victoria *Times Colonist* and *Vancouver Sun* Articles," The Land Conservancy, September 15, 2010, accessed June 21, 2014, http://blog.conservancy.bc.ca/2010/09/tlcs-qa-from-victoria-times-colonist-and-vancouver-sun-articles/.

332 "Bill Turner, TLC, interview," The Land Conservancy, accessed June 21, 2014, http://blog.conservancy.bc.ca/wp-content/uploads/2011/06/Bill-Turner-TLC-Interview.mp3.

333 "TLC's Q&A."

334 The Land Conservancy's five regional areas are: the Kootenay Region, the Lower Mainland Region, the Northern Region, the Okanagan Region and the Vancouver Island Region, as detailed in "Properties," The Land Conservancy, accessed August 20, 2014, http://blog.conservancy.bc.ca/properties/.

335 "Strange Doings at The Land Conservancy," Club Tread, May 19, 2009, http://www.clubtread.com/sforum/topic.asp?TOPIC_ID=30707.

336 "Letter to the Editor from TLC Board to Victoria *Times Colonist*," The Land Conservancy, September 21, 2011, http://blog.conservancy.bc.ca/2011/09/letter-to-the-editor-from-tlc-board-to-victoria-times-colonist/.

337 "TLC's Q&A."

338 "FAQ's from TLC's Board of Directors," The Land Conservancy, November 22, 2012, http://blog.conservancy.bc.ca/2012/11/faqs-from-tlcs-board-of-directors/; and "Update to TLC Members from John Shields," September 4, 2014, http://

blog.conservancy.bc.ca/2014/09/ update-to-tlc-members-from-john-shields/. For more detailed information on the court cases, results of appeals in the Binning House, properties listed for sale, reports to the court, judgments from the court, media articles and other issues involved in the Companies' Creditors Arrangement Act (CCAA) – a Canadian federal act that allows financially troubled corporations the opportunity to restructure their affairs – see TLC's blog "Latest CCAA Updates," http://blog.conservancy.bc.ca/ latest-updates/.

339 Bill Turner interview with Robin Alys Roberts, Spring 2014. See also, Amanda Hansen, "The Man Behind TLC: Founder Bill Turner," The Land Conservancy, Fall 2012, http://blog. conservancy.bc.ca/wp-content/ uploads/2012/10/landmark-fall-2012-low-res.pdf.

340 See http://internationaltrusts. org/.

341 "TLC Executive Director Receives 2012 Queen Elizabeth Diamond Jubilee Medal," The Land Conservancy, March 19, 2012, http://blog.conservancy. bc.ca/2012/03/tlc-executive-director-receives-2012-queen-elizabeth-diamond-jubilee-medal/.

342 Bill Turner interview with Robin Alys Roberts.

343 Ibid.

344 "Dr. Richard Hebda, Curator of Botany and Earth History, Royal BC Museum," Royal BC Museum, accessed June 22, 2014, http://dinosaurs. royalbcmuseum.bc.ca/

wp-content/uploads/2012/04/ Backgrounder-Dr.-Richard-Hebda-Curator-Royal-BC-Museum.pdf.

345 "Bob Maxwell," Soil 4 Youth, accessed June 22, 2014, http:// soilweb.landfood.ubc.ca/youth/ career-path/meet-the-pros/ bob-maxwell.

346 Amy Reiswig, "Ancient Pathways, Ancestral Knowledge: Nancy Turner's Monumental New Work Explores Humanity's Multi-Faceted Relationship to Plants," Focus Online, June 2014, http://www.focusonline. ca/?q=node/729.

347 Find out more about Woodwynn Farms at http:// www.woodwynnfarms.org/.

348 "Feast of Fields," FarmFolk CityFolk, accessed June 22, 2014, http://www.farmfolkcityfolk.ca/ events/feast-of-fields/.

349 "Event History," FarmFolk CityFolk, accessed June 22, 2014, http://www.farmfolkcityfolk. ca/events/feast-of-fields/ event-history/.

350 Matthew Robinson, "Jimmy Pattison's $9.5 billion Make Him the Richest Man in Canada," The Vancouver Sun, December 17, 2013, http://www. vancouversun.com/business/ Jimmy+Pattison+billion+mak es+richest+Canada/9298466/ story.html.

351 "The Ed Johnston [sic] Challenge," YouTube Video, 0:35, at "The Race Is On to Protect Madrona," The Land Conservancy, January 11, 2009, http://blog.conservancy. bc.ca/2009/01/the-race-is-on-to-protect-madrona/.

352 "Bluffs Park," Gulf Islands Eh,
accessed March 16, 2015, http://
www.gulfislandseh.com/galiano/
bluff.htm.

Eight: Choose a Model

353 Reinhold Niebuhr was an
American theologian and ethicist
from Wright City, Missouri. See
http://thinkexist.com/quotation/
change_is_the_essence_of_
life-be_willing_to/254730.html.

354 See chapter 11 in this book.

355 Edward Hill, "Saanich Tree
Protection Bylaw Unfair, Say
Homebuilders," *Victoria News*,
August 13, 2013, http://www.
vicnews.com/news/219424971.
html.

356 See http://www.fncv.org.au/
vicnat.htm.

357 See http://www.foodroots.ca/.

358 See http://www.shareorganics
bc.ca/.

359 See http://lifecyclesproject.ca/.

360 See http://www.farmlandstrust.
ca/.

361 See http://penderIslandweb.com/
farm/.

362 See http://www.
horselakefarmcoop.ca/.

363 See American Farmlands Trust,
http://www.farmland.org/.

364 "We Are Proud Members of the
Conservation Partners Program,"
Happy Valley Lavender Farm,
accessed August 20, 2014, http://
www.happyvalleylavender.com/
tlc-partner.php.

365 "Welcome to the National Trust
for Land and Culture," National
Trust for Land and Culture,
accessed June 22, 2014, http://ntlc.
ca/.

366 "International Trusts
Organisation: 2009 Annual
Report and Accounts,"
International Trusts
Organisation, accessed
August 20, 2014, http://
internationaltrusts.org/
wp-content/uploads/2013/08/
INTO-Annual-Report-2009-
for-website.pdf.

367 "The Hutchison Story," Saanich,
accessed June 22, 2014, http://
www.saanich.ca/parkrec/parks/
documents/BruceHutichsonPark
HutchisonFamilyHistory.pdf.
[Typo forms part of URL.]

368 "Planning for Profit, Five Acre
Mixed Vegetable Operation:
Full Production, Vancouver
Island, Spring 2008,"
British Columbia Ministry
of Agriculture and Lands,
accessed March 16, 2015, http://
www.agf.gov.bc.ca/busmgmt/
budgets/budget_pdf/small_
scale/2008mixed%20veg.pdf.
Another helpful cost reference
can be found in appendices A
(spreadsheets and crop value
charts) and B (Sample Sales
Spreadsheet, Sample Weekly
Invoice and Income Statement),
in Rachel Fisher, Heather
Stretch and Robin Tunnicliffe,
*All the Dirt: Reflections on
Organic Farming* (Victoria:
TouchWood Editions, 2012).

369 "Community Supported
Agriculture C.S.A., Sudoa Farm
in British Columbia," City
Farmer, accessed June 23, 2014,
http://www.cityfarmer.org/csa.
html.

370 "British Columbia," Know
Where Your Food Comes From,
accessed June 23, 2014, http://

knowwhereyourfoodcomesfrom.
com/community-supported-
agriculture-csa-farms/canada/
british-columbia/.

371 See http://www.
farmfolkcityfolk.ca/.

372 See http://thewholebeast.ca/.

373 See www.alderleafarm.
com/#!farm/viewstack3=csa.

374 See http://www.cityfarmscoop.
ca/.

375 See http://www.foxglovefarmbc.
ca/.

376 See www.freshroots.ca.

377 See http://
glenvalleyorganicfarmcoop.
org/coop/.

378 Axiom News, "Farm Co-op
Seeds Resilient Communities,"
British Columbia Co-operative
Association, January 22, 2014,
http://bcca.coop/news/farm-
co-op-seeds-resilient-
communities. See also, "Fraser
Common Farm Cooperative,
Aldergrove," The Land
Conservancy, accessed June 23,
2014, http://blog.conservancy.
bc.ca/get-involved-with-tlc/
resources/conservation-
partners-
program/our-partners-lower-
mainland-region/fraser-
common-farm-cooperative-
aldergrove/.

379 See www.greencityacres.com.

380 See http://haliburtonfarm.org/
wp/.

381 See http://www.
horselakefarmcoop.ca/.

382 See http://www.
horselakefarmcoop.ca/ceeds/.

383 See http://www.lawnstolegumes.
ca.

384 See http://www.makariafarm.
com/.

385 See http://www.nathancreek.
ca/.

386 See http://notchhillorganic.
ca/. Notch Hill part of a
long list of organic farms
that provide delivery in the
Central Okanagan area of BC:
"Suppliers," Urban Harvest,
accessed March 16, 2015, http://
urbanharvest.ca/suppliers/.

387 See http://yecfarm.blogspot.ca/.

388 See http://ourecovillage.org/.

389 See http://providence.bc.ca/.

390 See http://soilmatters.
wordpress.com/.

391 See http://www.landfood.ubc.
ca/ubcfarm/csa.php.

392 See, http://www.uvic.ca/
research/centres/cccbe/
resources/galleria/stories/
VancouverIslandOrganicCoop.
php.

393 European Commission, "The
Common Agricultural Policy:
A Partnership between Europe
and Farmers" (Luxembourg:
Publications Office of the
European Union, 2012), http://
ec.europa.eu/agriculture/cap-
overview/2012_en.pdf. See also,
"Overview of CAP Reform
2014–2020," Agricultural
Policy Perspectives Brief, no.
5 (December 2013), http://
ec.europa.eu/agriculture/
policy-perspectives/
policy-briefs/05_en.pdf.

394 Eva Antczak, "California
Farmland Protection: Reality
or Wishful Thinking?"
FarmsReach (blog), August 8,
2013, http://blog.farmsreach.
com/california-farmland-

protection-reality-or-wishful-thinking/.

395 European Commission, "The Common Agricultural Policy."

396 Ibid.

397 Ibid.

398 Ibid.

399 L.V. Dicks, J.E. Ashpole, J. Danhardt, K. James, A. Jönsson, N. Randall, D.A. Showler, R.K. Smith, S. Turpie, D. Williams and W.J. Sutherland, *Farmland Conservation: Evidence for the Effects of Interventions in Northern and Western Europe, Synopses of Conservation Evidence, Volume 3* (Exeter, UK: Pelagic Publishing, 2013), www.conservationevidence.com/synopsis/download/9.

400 Stew Hilts, "Saving the Land That Feeds Us: How to Revitalize Our Near-urban Farmland and Curb Sprawl," *Alternatives Journal*, April 2008, http://www.alternativesjournal.ca/sustainable-living/saving-land-feeds-us.

401 Antczak, "California Farmland Preservation."

402 Ibid.

403 "We are PCC Farmland Trust," PCC Farmland Trust, accessed August 21, 2014, http://www.pccfarmlandtrust.org/.

404 "Harvesting Opportunities in New York, 2013 Conference Inspires and Educates," American Farmland Trust, New York State Office, accessed June 15, 2014, http://newyork.farmland.org/harvestingopportunities2013/.

405 "Annam Project: The Good Food Movement in Kerala," The Environment Collaborative, accessed June 25, 2014, http://environmentcollaborative.org/our-work/anna-swaraj/annam.

406 "Annam – Good Food Movement," Centre for Innovation in Science and Social Action, accessed August 21, 2014, http://www.cissa.co.in/index.php/programmes/13-annam-good-food-movement/event_details.

407 "Permaculture Fundamentals," Open Permaculture School, accessed June 25, 2014, http://www.permaculturedesigntraining.com/category/permaculture-fundamentals.

408 See http://www.wwoof.net/.

409 See http://wwoofcsa.com/wwoofcsa/.

410 Ibid.

Nine: Identify Vital Farmland

411 Lee M. Talbot, "A World Conservation Strategy: The Inaugural American Exchange Lecture," *Journal of the Royal Society of Arts* 128, no. 5288 (July 1980): 493–510, http://www.jstor.org/discover/10.2307/41373126?uid=3739400&uid=2&uid=3737720&uid=4&sid=21104949447807.

412 "Madrona: A Precious Farm in Greater Victoria," The Land Conservancy, accessed June 25, 2014, http://blog.conservancy.bc.ca/wp-content/uploads/2009/12/MadronaFarmProspectus.pdf.

413 A. Wezel, S. Bellon, T. Doré, C. Francis, D. Vallod and D. David, "Agroecology as a Science, a Movement and a Practice,"

Agroecology in Action, 2009, accessed June 25, 2014, http://agroeco.org/socla/wp-content/uploads/2013/12/wezel-agroecology.pdf.

414 "Farmer David Humane Bullfrog Hunting," YouTube video, 0:46, posted by "David Chambers," June 27, 2011, http://www.youtube.com/watch?v=RskGoQCFraQ.

415 "Invasive Species Management," Madrona Farm, accessed June 25, 2014, http://madronafarm.co/site.php?rsm=1&action=projects&blabber_buster=blab_5&o=surpress_ads&1=true#5.

416 "Agriculture Capability," Agricultural Land Commission, accessed June 25, 2014, http://www.alc.gov.bc.ca/alr/Ag_Capability.htm.

417 Melissa Mayntz, "North American Migration Flyways," About.com, accessed June 25, 2014, http://birding.about.com/od/birdingbasics/ss/North-America-Migration-Flyways.htm.

418 "Priority Birds," Audubon, accessed June 25, 2014, http://conservation.audubon.org/priority-birds.

419 "Local Food and the Environment: Sustainable Agriculture, Environmental Stewardship," Coquitlam Farmers Market: Make, Bake or Grow, accessed August 20, 2014, http://www.makebakegrow.com/market-news/local-food-and-the-environment-sustainable-agriculture-environmental-stewardship/.

420 "Harold Steves, the 'Father of the ALR,'" YouTube video, 32:04, posted by "peninsula video," December 1, 2013, http://www.youtube.com/watch?v=Al_3kfOpOe4.

Ten: Build Community

421 Aldo Leopold, *A Sand County Almanac* (New York: Oxford University Press, 1968).

422 Ben Isitt, "Langford's Bear Mountain Interchange: Urbanization on the Western Frontier and the Blurring of Public and Private Interests," First Nations, December 2007, http://www.firstnations.de/media/06-0-isitt.pdf.

423 "Biography," Story Clark, accessed June 26, 2014, http://www.storyclark.net/bio.htm.

424 Story Clark, "Coping with Crisis: Strategies for Helping Land Trusts through the Recession," *Saving Land* (Winter 2009): 17, http://www.storyclark.net/files/SavingLand-coverstory-clark.pdf.

425 "A Field Guide to Conservation Finance," Story Clark, praise by Terry Tempest Williams, accessed June 26, 2014, http://www.storyclark.net/_center__i_a_field_guide_to_conservation_finance__i___center__64775.htm.

426 "What Kind of World Do You Want," Jim Lord, accessed June 26, 2014, http://www.jimlord.org/world/what-kind-of-world-do-you-want.

427 David Cooperrider and

Diana Whitney, "What Is Appreciative Inquiry?" Appreciative Inquiry Commons, accessed June 26, 2014, http://appreciativeinquiry.case.edu/intro/whatisai.cfm.

428 "Food and Farming," National Trust, accessed August 20, 2014, https://www.nationaltrust.org.uk/what-we-do/big-issues/food-and-farming/what-were-doing/fine-farm-produce-awards/?campid=Print_Central_FFPA2014.

429 Since they first forged partnerships with Madrona, some of the restaurants in the following list have closed and/or chefs have moved on, some to places farther north on Vancouver Island. However, many of these restaurants and chefs are still well-known and loved around Victoria, and the referenced links still provide educational information that may prove helpful wherever one lives:

Brasserie L'ecole (see http://www.lecole.ca/). Top-rated head Chef Sean Brennan "joined the effort because he feels it is important for chefs to be part of the circle creating awareness." (See "The Madrona Farm Chef Survival Challenge," *Eat*, http://eatmagazine.ca/the-madrona-farm-chef-survival-challenge/).

Caffé Fantastico (see http://www.caffefantastico.com/).

Camille's, Chef Patrick Miller (see http://www.camillesrestaurant.com/).

Fairmont Empress, Chef Ken Nakano (See "Dining," The Fairmont Empress, accessed June 27, 2014, http://www.fairmont.com/empress-victoria/dining/.)

La Piola (now closed), Chef Alberto Pozzolo (see https://www.facebook.com/LaPiolaGastronomia). Also Chef Cory Pelan ("How LaPiola Restaurant Donates to Protect Madrona Farm," YouTube video, 1:49, posted by "David Chambers," November 6, 2013, http://www.youtube.com/watch?v=jUDtYJuWfm4&hd=1).

Lure, Chef Michael Weaver (see https://www.facebook.com/Lureseafoodandgrille).

Pacific Prime Restaurant, Chef Stephan Drolet (see http://www.beachclubbc.com/amenities/dining/).

Pizzeria Prima Strada, Chef Jonathan Pulker (see http://pizzeriaprimastrada.com/).

Royal Colwood Golf Course, Chef Chris Hammer (see https://www.royalcolwood.org/The-Club/Executive.aspx).

Smoken Bones (now closed), Chef John Brooks (Chef John Pulker) (see David Lennam, "Ken Hueston, Smoken Bones Cookshack," *Douglas Business Magazine*, accessed March 16, 2015, http://douglasmagazine.com/resources/42-profiles/112-ken-hueston.html. See also Carla Wilson, "Doors Are Closed at Smoken Bones in Victoria's Hudson," *Times Colonist*, June 25, 2014, http://

www.timescolonist.com/
business/doors-are-closed-at-
smoken-bones-in-victoria-s-
hudson-1.1158715).

Solomon's (now closed), Chef
Michael DeGrazia (see http://
vibrantvictoria.ca/local-
news/victorias-cocktail-bar-
pioneer-closes-doors-after-
18-months-in-business/ and
http://tasteandsipmagazine.
com/2011/10/25/interview-with-
solomon-siegel/).

The Italian Bakery, Chef
Alberto Pozzolo (see "Who We
Are," Ottavio Italian Bakery
and Delicatession, accessed
March 16, 2015, http://www.
ottaviovictoria.com/whoweare.
php).

The Whole Beast, owner Cory
Pelan (see http://thewholebeast.
ca/ and "Cory Pelan
Profile – Slow Food Canadian
Food Heroes," YouTube video,
3:35, posted by "FeastTV,"
https://www.youtube.com/
watch?v=fEgZQwdNcF0).

Vista 18, Chef Mike Dunlop
(see http://www.vista18.com/).

Zambri's (see http://www.
zambris.ca/), Chef Peter
Zambri (see Grania Litwin,
"Chefs Take to the Wilds in
Foodie Survival Test," *Times
Colonist*, October 2, 2009,
http://www.timescolonist.
com/life/chefs-take-to-the-
wilds-in-foodie-survival-test-
1.11827#sthash.pJ9RJroi.dpuf).

430 Paul Waldie, "A Century of
Good Deeds," *The Globe and
Mail*, November 19, 2009,
http://www.theglobeandmail.
com/report-on-business/a-
century-of-good-deeds/
article4292614/.

431 "The Ed Johnston [sic]
$200,000 Challenge," YouTube
video, 0:35 posted by "David
Chambers," November 6, 2013,
http://www.youtube.com/
watch?v=11L9CBuvWOY.

432 *EcoNews* is now called
EarthFuture and is available at
http:www.earthfuture.com/.

433 See https://www.facebook.
com/groups/46722105836/.

434 See http://lifecyclesproject.ca/.

435 See http://mustardseed.ca/.

436 "Projects," Lifecycles,
accessed June 27, 2014, http://
lifecyclesproject.ca/initiatives/.

437 See http://www.growarow.org/.

438 See http://ourplacesociety.
com/.

439 "Growing Schools," Lifecycles,
accessed June 28, 2014, http://
lifecyclesproject.ca/initiatives/
growing_schools/.

440 See http://
thepollinationproject.org/.

441 See http://www.
sharingbackyards.com/.

442 See http://
urbanagriculturehub.ca/.

443 "Urban Agriculture,"
Lifecycles, accessed June 27,
2014, http://lifecyclesproject.
ca/index.php?action=
au_hub.

444 See http://www.gardenworks.
ca/.

445 "Giving Tree Program Gives
Trees to Madrona Farm,"
The Land Conservancy,
February 28, 2011, http://blog.
conservancy.bc.ca/2011/02/
giving-tree-program-gives-
trees-to-madrona-farm/.

446 "How It Works," Eat Local

Grown, accessed March 16, 2015, http://eatlocalgrown.com/how-it-works.html.

447 See http://www.ffcf.bc.ca/.

448 "Who We Are," FarmFolk CityFolk, accessed July 12, 2014, http://www.farmfolkcityfolk.ca/who-we-are/.

449 "Feast of Fields," FarmFolk CityFolk, accessed July 12, 2014, http://www.farmfolkcityfolk.ca/events/feast-of-fields/.

450 Links for these and other projects are available at "Projects," FarmFolk CityFolk, accessed March 16, 2015, http://www.farmfolkcityfolk.ca/projects/.

451 See http://www.iccbc.ca/about/members.

452 See http://transitionvictoria.ning.com/.

453 "Mission and Vision," Transition Victoria, accessed July 12, 2014, http://transitionvictoria.ning.com/page/mission-vision.

454 "Flax to Linen," Transition Victoria, accessed July 12, 2014, http://transitionvictoria.ning.com/page/the-linen-project.

455 "Transition Streets," Transition Victoria, accessed July 12, 2014, http://transitionvictoria.ning.com/page/transition-streets.

456 "Capital Nut Project," Transition Victoria, accessed July 12, 2014, http://transitionvictoria.ning.com/group/capital-nut-project.

457 "Springridge Common Map," Fernwood NRG, accessed October 1, 2014, http://fernwoodnrg.ca/fernwood-nrg-programs/urban-sustainability/spring-ridge-common/springridge-common-map/.

458 "Home Groups," Transition Victoria, accessed July 12, 2014, http://transitionvictoria.ning.com/page/home-groups.

459 Victoria Forest Action Network (VIC FAN), accessed December 20, 2014, http://forestaction.wikidot.com/

Eleven: Fundraise, Fundraise, Fundraise

460 Hank Rosso, *Achieving Excellence in Fundraising* (San Francisco: Jossey-Bass, 2011), 4, 70, 272, 362 and 374. Hank Rosso is the founder of The Fund Raising School. See "The Fund Raising School," Lilly Family School of Philanthropy, accessed March 16, 2015, http://www.philanthropy.iupui.edu/the-fund-raising-school.

461 "Madrona Farm," Heritage Canada, The National Trust, accessed July 12, 2014, http://www.heritagecanada.org/en/visit-discover/find-historic-places/historic-sites-canada/british-columbia/madrona.

462 "$287,000 to Go in the Race to Save Madrona Farm for The Land Conservancy," The Land Conservancy, accessed August 20, 2014, http://blog.conservancy.bc.ca/2010/01/287000-to-go-in-the-race-to-save-madrona-farm-for-the-land-conservancy/.

463 Grania Litwin, "Farm

Deadline Looms: Two Weeks Left to Raise Money for Madrona Purchase," *Times Colonist*, December 24, 2009.

464 Paul Waldie, "A Century of Good Deeds," *The Globe and Mail*, November 14, 2009, http://www.theglobeandmail.com/report-on-business/a-century-of-good-deeds/article4292614/.

465 Ibid.

466 Bill Tieleman, "Basi-Virk: Courtroom Confusion Explained," *Bill Tieleman* (blog), December 29, 2008, http://billtieleman.blogspot.ca/2008/12/a-z-of-basi-virkbc-legislature-raid.html

467 "The Green Interview: Vandana Shiva," The Green Interview, accessed October 20, 2014, http://www.thegreeninterview.com/interviews/vandana-shiva.

468 "Agricultural Land Reserve," Create Garden City Lands, accessed July 12, 2014, http://creategardencitylands.ca/agricultural-land-reserve/.

469 "Biodiversity," Food and Agricultural Organization of the UN, accessed December 20, 2014, http://www.fao.org/biodiversity/en/. "Biodiversity is essential for food security and nutrition. Thousands of interconnected species make up a vital web of biodiversity within the ecosystems upon which global food production depends."

470 "His Royal Highness, Charles, Prince of Wales on Food Security," YouTube video, 7:52, posted by "Council on

Foundations," April 8, 2013, http://www.youtube.com/watch?v=HlcOae-MBS8.

471 "Minutes of the Special Committee of the Whole Meeting Held in the Council Chambers, Saanich Municipal Hall," Saanich, September 16, 2008, http://www.saanich.ca/living/mayor/pdf/mins/2008/sep16spcwminutes.pdf.

472 "Urban Containment Boundary" (including Gordon Head) as of September 2013, Saanich, accessed March 16, 2015, http://saanich.ca/living/afs/docs/farm_map_jul_2011.pdf.

473 "Breaking News: Changes to ALR Could See More Business on Farmland," *The Province*, March 27, 2014, http://blogs.theprovince.com/2014/03/27/breaking-news-changes-to-alr-could-see-more-business-on-farmland/.

474 "Interview: ALR Changes Help Farmers, Farm Food Security," *The Province*, April 1, 2014, http://blogs.theprovince.com/2014/04/01/interview-alr-changes-help-farmers-hurt-food-security/.

475 Bill Cleverley, "Saanich Mulls Fate of Cannons," *Times Colonist*, September 11, 2012, http://www.timescolonist.com/news/local/saanich-mulls-fate-of-cannons-1.14252.

476 Bill Cleverley, "Saanich Scoops Blenkinsop Lake and Land Ringing Farm," *Times Colonist*, September 27, 2013, http://www.timescolonist.com/news/local/saanich-

scoops-blenkinsop-lake-and-land-ringing-farm-1.641706.

477 "Saanich Council Briefs: Plans, Parks and Purchases … Municipality Buys Land Next to Mt. Doug Park," *Saanich News*, May 19, 2011, http://www.saanichnews.com/news/122247124.html. Also see "TLC's Q&A from Victoria *Times Colonist* and *Vancouver Sun*," The Land Conservancy, September 15, 2010, http://blog.conservancy.bc.ca/2010/09/tlcs-qa-from-victoria-times-colonist-and-vancouver-sun-articles/.

478 Robert Randall, "Madrona Farm Buyout Nears Deadline," *Vibrant Victoria*, March 4, 2010, http://vibrantvictoria.ca/local-news/madrona-farm-buyout-nears-deadline/.

479 "Family Farm Seeks Future with TLC," *Times Colonist*, June 19, 2008, http://www.canada.com/victoriatimescolonist/news/life/story.html?id=78bfae85-25ab-4f78-b3fe-ce58d7168fc9.

480 "The Roadside Chef: Saanich Farmer Attracts a Crowd to His Rural Cooking Demonstrations," *Times Colonist*, September 23, 2006, http://www.canada.com/story.html?id=8834ab75-31a4-4539-98eb-d84aebb953ef.

481 Grania Litwin, "Chefs Take to the Wilds in Foodie Survival Test," *Times Colonist*, October 2, 2009, http://www.timescolonist.com/life/chefs-take-to-the-wilds-in-foodie-survival-test-1.11827.

482 "Madrona Farms Blackberry," Phillips Beer, accessed July 12, 2014, http://phillipsbeer.com/madrona-farms-blackberry.

483 "Feast of Fields, September 19 on Vancouver Island," The Land Conservancy, September 9, 2010, http://blog.conservancy.bc.ca/2010/09/feast-of-fields-september-19-on-vancouver-island/.

484 "Local Food Micro-Loan Fund," FarmFolk CityFolk, accessed August 20, 2014, http://www.farmfolkcityfolk.ca/resources/microloan/.

485 "PSA: Rock & Royal Tea Party: A Dinner for Madrona Farm," The Land Conservancy, January 16, 2009, http://blog.conservancy.bc.ca/2009/01/psa-rock-a-dinner-for-madrona-farm/.

486 "Save Madrona Farm with $1," YouTube video, 1:31, posted by "David Chambers," May 25, 2009, http://www.youtube.com/watch?v=tU9AUimMFCo.

487 Christine van Reeuwyk, "Woodwynn Farm Sold for Homeless Solution," Woodwynn Farms, March 18, 2009, http://www.woodwynnfarms.org/2009/03/18/woodwynn-farms-in-the-peninsula-news-review-march-18-2009/.

488 "Funniest Madrona Farm Christmas, Give the Gift of Food Security," YouTube video, 1:35, posted by "David Chambers," November 16, 2009, http://www.youtube.com/watch?v=APNYcHl1NzU.

489 Erika Verlinden, "Youth

Help Plant the Final Seeds of Support for Madrona Farm," The Land Conservancy, accessed July 12, 2014, http://blog.conservancy.bc.ca/wp-content/uploads/2010/03/LandMark-March-2010-low-res.pdf.

490 "The Big Dream Farm Fund, Great News for Farmland Conservation!" Chef Survival Challenge, accessed December 8, 2014, http://chefsurvivalchallenge.ca/?page_id=18. See also Gurdeep Stephens, "Big Dream Farm Fund," *The Deeper Side* (blog), accessed December 8, 2014, http://thedeepersideblog.com/2013/05/31/big-dream-farm-fund/.

491 See http://www.fncv.org.au/the-zzzzzzzzzzzz-naturalist/.

492 See http://www.foodroots.ca/.

493 See http://www.shareorganics.bc.ca/.

494 See http://lifecyclesproject.ca/.

495 See http://www.farmlandstrust.ca/.

496 See http://penderislandweb.com/farm/.

497 See http://www.horselakefarmcoop.ca/.

498 "The Big Dream Farm Fund," Chef Survival Challenge, accessed July 12, 2014, http://chefsurvivalchallenge.ca/?page_id=18.

499 "Madrona Farm,' Chef Survival Challenge, accessed October 20, 2014, http://chefsurvivalchallenge.ca/?page_id=16.

500 Ibid.

Twelve: Trust the Commons

501 Frances Moore Lappé, "Commons Care: How Wrong Was Garrett Hardin?!" *Huffington Post,* April 9, 2013, http://www.huffingtonpost.com/frances-moore-lappe/commons-care-how-wrong_b_3039549.html.

502 See http://cityrepair.org/.

503 See http://www.communitecture.net/.

504 Peter Barnes, *Capitalism 3.0: A Guide to Reclaiming the Commons* (San Francisco: Berrett-Koehler Publishers, Inc., 2006), 87–88. A free, electronic version is available at http://www.onthecommons.org.

505 "About," Remix the Commons, accessed July 12, 2014, http://www.remixthecommons.org/en/a-propos/. "Remix the Commons was initiated in 2010 by Alain Ambrosi with the organizations of Communautique and VECAM, in Québec and France."

506 Barnes, *Capitalism 3.0*, 87–88.

507 Ariana McBridge, "Frances Moore Lappé's *Getting A Grip: Creativity and Courage in a World Gone Mad*," Orton Family Foundation, Summer 2008, http://www.orton.org/resources/publications/scenarios/scenarios_e_journal/frances_moore_lappe.

508 "99 Reasons Why It's Better to Be Canadian: Our Sixth Annual Canada Day Survey,"

Maclean's, June 28, 2013, http://www.macleans.ca/news/canada/99-reasons-why-its-better-to-be-canadian/.

509 Barnes, *Capitalism 3.0*, 85–86.

510 "Founding Story," Marin Agriculture Land Trust (MALT), accessed July 12, 2014, http://www.malt.org/founding-story.

511 "Best & Wayburn Weigh 15 Years of Work: Conservation, Commerce, Partnerships and Persistence Define PFT," Pacific Forest Trust, *ForestLife* (Winter 2008): 4–5, https://pacificforest.org/best-and-wayburn-weigh-15-years-of-work.html.

512 Barnes, *Capitalism 3.0*, 85–88.

513 "Elinor Ostrom, Nobel Laureate," Indiana University, accessed July 12, 2014, http://elinorostrom.indiana.edu/. See also "Elinor Ostrom: Facts," Nobel Prize, accessed July 12, 2014, http://www.nobelprize.org/nobel_prizes/economic-sciences/laureates/2009/ostrom-facts.html.

514 Frances Moore Lappé, *EcoMind: Changing the Way We Think, to Create the World We Want* (Cambridge, Mass.: Small Planet Institute, 2011), 127.

515 James S. Fishkin, *When the People Speak* (Oxford: Oxford University Press, 2009), 124.

516 Lappé, *EcoMind*, 127.

517 As well as often being the first plant to pop up after a forest fire, the lovely magenta fireweed has many health benefits for humans. "Fireweed," Herbs 2000, accessed July 12, 2014, http://www.herbs2000.com/herbs/herbs_fireweed.htm.

518 Lappé, *EcoMind*, 46, 123 and 253. See also Miguel Mendonca, David Jacobs and Benjamin Sovacool, *Powering the Green Economy: The Feed-in Tariff Handbook* (London: Earthscan, 2009).

519 Lappé, *EcoMind*, 123–24.

520 Ibid., 122–23.

521 See http://www.listeningproject.info/index.php.

522 Herb Walters, "Adventures in Listening," Quaker Universalist Fellowship, accessed July 12, 2014, http://universalistfriends.org/walters.html.

523 Lappé, *EcoMind*, 195–209.

524 Helen Forsey, "Standoff at Meyers Farm: Citizens and Social Media Are Pitted against the Special Ops," *Canadian Centre for Policy Alternatives (CCPA) Monitor* 20, no. 10 (April 2014): 22. See also http://www.policyalternatives.ca.

525 Erika Tucker, "Demolition to Begin on Frank Meyers' Farm amid Claims of Government Bullying," *Global News*, January 13, 2014, http://globalnews.ca/news/1079592/demolition-to-begin-on-frank-meyers-farm-amid-claims-of-government-bullying/.

526 Michael Friscolanti, "Canadians Join 85-Year-Old Frank Meyers in Battle to Save Farm. But Is All the Attention Too Little, Too Late?" *Maclean's*, October 13,

2013, http://www.macleans.ca/
news/canada/canadians-join-
85-year-old-frank-meyers-in-
battle-to-save-farm/.

527   Forsey, "Standoff," 22.

528   Save Frank and Marjorie
      Meyers Farm Facebook page,
      accessed August 14, 2014,
      https://www.facebook.com/
      pages/Save-Frank-Marjorie-
      Meyers-Farm/
      1419959788219689?ref=
      br_tf.

529   Ibid.

530   "Historica Canada, Crown
      Land," *The Canadian
      Encyclopedia*, accessed
      August 20, 2014, http://www.
      thecanadianencyclopedia.
      ca/en/article/crown-land/.
      "Crown land is the term
      used to describe land owned
      by the federal or provincial
      governments. Authority for
      control of these public lands
      rests with the Crown, hence
      their name. Less than 11% of
      Canada's land is in private
      hands; 41% is federal crown
      land and 48% is provincial
      crown land."

531   Heather Menzies, "Commons
      of the Past Inspiring Model
      for Today's Progressives,"
      *Canadian Centre for Policy
      Alternatives (CCPA) Monitor*
      20, no. 10 (April 2014), 1, 6.

532   Heather Menzies, *Reclaiming
      the Commons for the Common
      Good: A Memoir and
      Manifesto* (Gabriola Island,
      BC: New Society Publishers,
      2014), ix.

533   Ibid., 69.

534   Ibid., 68–71.

535   See http://www.occupy.net/

and http://onthecommons.
org/occupy-commons; and
also http://onthecommons.
org/work/commons-network/
join-commons-network.

536   See http://www.idlenomore.
      ca/, and Idle No More
      Community Facebook page,
      accessed August 20, 2014,
      https://www.facebook.com/
      IdleNoMoreCommunity.
      See also https://twitter.com/
      search?q=%23idlenomore.

537   See http://www.
      communitecture.net/.

538   See http://ocw.mit.edu/index.
      htm.

539   See http://oyc.yale.edu/.

540   See http://www.outreach.
      washington.edu/openuw/.

541   See https://oli.cmu.edu/.

542   "Free Online Education
      Courses," Guide to Online
      Schools, accessed July
      12, 2014, http://www.
      guidetoonlineschools.com/
      articles/getting-started/
      free-online-classes.

543   See http://saltspringbuilder.
      com/.

544   For a clear definition of
      co-counselling, and the
      principles of how the
      equal sharing of client and
      counsellor positions works,
      see J. Heron, "A Definition of
      Co-counselling International
      (CCI)," CoCo Info, 1996,
      accessed July 12, 2014, http://
      co-counselling.info/en/ref/
      definition-co-counselling-
      international-cci.

545   "Welcome to WWOOF,"
      World Wide Opportunities on
      Organic Farms, accessed July

12, 2014, http://www.wwoof.net/welcome-to-wwoof//.

546 Menzies, *Reclaiming the Commons*, 184.

547 "Mushroom Volunteer Day at Madrona Farm," The Land Conservancy, April 11, 2011, http://blog.conservancy.bc.ca/2011/04/mushroom-volunteer-day-at-madrona-farm/.

Thirteen: Share Mutual Success

548 See http://izquotes.com/quote/54945. Born into an Italian family, a graduate in anthropology from UCLA and a graduate chef from Le Cordon Bleu in Paris, Giada de Laurentiis has written several enticing cookbooks and is a featured TV personality. Her website is http://www.giadadelaurentiis.com/.

549 Jeff Nield, "Arran Stephens of Nature's Path on Independence in the Fast-Consolidating Food Industry," *Treehugger*, October 31, 2008, http://www.treehugger.com/culture/arran-stephens-of-natures-path-on-independence-in-the-fast-consolidating-food-industry.html.

550 Steve Burgess, "Food for Thought," *Nuvo Magazine*, Autumn 2013, http://nuvomagazine.com/magazine/autumn-2013/natures-path.

551 "Organic Cereal Profits Naturally," Arran Stephens interview, *Trust Organic Food*, accessed March 10, 2013, http://trustorganicfood.com/organic-cereal-profits-naturally/.

552 As explained at http://www.sos.org/can/western/page/meditation-ecology-centre-richmond-bc.html, "The SOS Meditation & Ecology Centre is dedicated to inner and outer peace through meditation and to caring for our internal and external environments. . . . we offer classes in organic gardening, vegetarian cooking and healthy living. There are organic vegetable and flower gardens, an orchard, bee hives and a demonstration of natural composting. The programs are always free."

553 Rupert Stephens (taken from a document typed by Ratana in 1977, the year after Rupert died). A slightly different version exists at "Humble Beginnings," Nature's Path, accessed July 12, 2014, http://ca-en.naturespath.com/blog/2009/03/28/humble-beginnings. His daughter sings this version of the song at: "This Earth Is Ours," Arran Stephens, accessed March 16, 2013, http://www.arranstephens.com/2013/07/this-earth-is-ours-song/.

554 "Harold Steves: BC's Patriarch of Farmland Preservation," YouTube video, 3:47, posted by "CommonSenseCanadian," April 20, 2011, http://www.youtube.com/watch?v=_J4e0gpg0yQ.

555 "#436 One Man's Vision Saved British Columbia's Farmland," This Gives Me Hope, September 25, 2012, http://thisgivesmehope.com/2012/09/25/436-one-mans-vision-saved-british-columbias-farmland/.

556 Sandor Gyarmati, "The ALR at 40," *Delta Optimist*, February 8, 2013, http://www.delta-optimist.com/news/the-alr-at-40-1.456781.

557 "How the ALR was Established," Agricultural Land Commission, accessed July 12, 2014, http://www.alc.gov.bc.ca/alr/Establishing_the_ALR.htm.

558 Gyarmati, "The ALR at 40."

559 "Councillor Harold Steves," Richmond, accessed July 12, 2014, http://www.richmond.ca/cityhall/council/about/members/steves.htm.

560 "Farm Facts or Fiction," Steveston Stock and Seed Farm, accessed July 12, 2014, http://www.stevesfarm.com/steveston/2011/10/24/farm-facts-or-fiction/.

561 "Farm Facts or Fiction," Pinnacle Farms, accessed July 12, 2014, http://www.pinnaclefarms.ca/FarmFactsOrFiction/FarmFactsOrFiction.html.

562 See http://www.eatwild.com/.

563 See http://www.seeds.ca/.

564 David Tracey, "Small Farmers: Vital Work, Slim Wages," *The Tyee*, September 1, 2009, http://thetyee.ca/News/2009/09/01/FarmerWages/.

565 David Tracey, "Why Urban Farming Is the Future and Why It's Good to Get a Little Dirty While Helping BC Feed Itself," *The Tyee*, August 18, 2009, http://thetyee.ca/News/2009/08/18/UrbanFarmingFuture/.

566 J.B. McKinnon, *The Once and Future World: Nature as It Was, as It Is, as It Could Be* (Toronto: Random House, 2013).

567 "Harold Steves: March against Monsanto – We Want to Know What's GMO," video posted at Vancouver Media Co-op, October 23, 2013, http://vancouver.mediacoop.ca/video/harold-steves-we-want-know-what%E2%80%99s-gmo-march-agains/19435.

568 Rod Mickleburgh, "Harold Steves' Unwavering Passion for the Land," *The Globe and Mail*, August 19 2013, http://www.theglobeandmail.com/news/british-columbia/harold-steves-unwavering-passion-for-the-land/article13835932/.

569 For a description of Harold Steves's farming and extensive political services, see "Councillor Harold Steves."

570 Rod Mickleburgh, "Land Reserve Was Brainchild of Disgruntled Farmer-Councillor," *The Globe and Mail*, September 17, 2012, http://www.theglobeandmail.com/news/british-columbia/land-reserve-was-brainchild-of-disgruntled-farmer-councillor/article4549133/.

571 See http://www.foodroots.ca/.

572 See http://www.shareorganics.bc.ca/.

573 See http://www.iccbc.ca/.

574 See chapter 10 for a complete list.

575 Jason Youmans, "Sowing the Seeds, Keeping Madrona Farm's Good Earth," *Monday*

Magazine, July 9–15, 2009.

576 "Welcome to the National Trust for Land and Culture (B.C.)" National Trust for Land and Culture, British Columbia, accessed July 12, 2014, http://ntlcbc.com/.

Fourteen: Keep Surging Forward

577 Michelangelo, "Famous Michelangelo Quotes," Michelangelo Gallery, accessed October 20, 2014, http://www.michelangelo-gallery.com/michelangelo-quotes.aspx. This quote is disputed but widely attributed to Michelangelo since the late 1990s.

578 Bill Turner is the founder of BC's The Land Conservancy (TLC) in 1997 and its long-time executive director. "Celebrating a TLC Legend: Bill Turner," The Land Conservancy, July 24, 2012, http://blog.conservancy.bc.ca/2012/07/celebrating-a-tlc-legend-bill-turner/.

579 See "Speaker Biographies: Bill Turner, The Land Conservancy of B.C.," Nanoose Naturalists, accessed October 20, 2014, http://www.nanoosenaturalists.org/speakers-and-outings-schedule/speakers-biographies/.

580 "Welcome to the National Trust for Land and Culture (B.C.)," National Trust for Land and Culture, British Columbia, accessed July 12, 2014, http://ntlcbc.com/.

581 "NTLC Vision and Mandate," National Trust for Land and Culture, British Columbia, accessed July 12, 2014, http://ntlcbc.com/?page_id=471.

582 "TLC Executive Director Receives 2012 Queen Elizabeth Diamond Jubilee Medal," The Land Conservancy, March 19, 2012, http://blog.conservancy.bc.ca/2012/03/tlc-executive-director-receives-2012-queen-elizabeth-diamond-jubilee-medal/. To see both Bill Turner's and TLC's impressive (and much longer) list of awards, see "Awards and Recognition," The Land Conservancy, accessed July 12, 2014, http://blog.conservancy.bc.ca/wp-content/uploads/2010/12/2010-Awards-and-Recognition.pdf.

583 See Nancy Turner's website at http://pspaldin.wix.com/nancyturner. See also her biography at "Turner, Nancy," University of Victoria, accessed March 16, 2015, http://www.uvic.ca/socialsciences/environmental/people/faculty/turnernancy.php.

584 Nancy Turner and Pamela Spalding, "'We Might Go Back to This': Drawing on the Past to Meet the Future in Northwestern North American Indigenous Communities," *Ecology and Society* 18, no.4 (2013): 29, http://dx.doi.org/10.5751/ES-05981-180429.

585 Nancy Turner and Patrick von Aderkas, "Sustained by First Nations: European Newcomers' Use of Indigenous Plant Foods in Temperate

North America," *Acta Societatis Botanicorum Poloniae* 81, no. 4 (December 2012): 295.

586 Nancy J. Turner, Fikret Berkes, Janet Stephenson and Jonathan Dick, "Blundering Intruders: Multi-Scale Impacts on Indigenous Food Systems," *Human Ecology* 41, no. 4 (August 2013): 563–74.

587 Nancy Turner, Colleen Robinson, Gideon Robinson and Belle Eaton, "'To Feed All the People': Lucille Clifton's Fall Feasts for the Gitga'at Community of Hartley Bay, British Columbia," in *Explorations in Ethnobiology: The Legacy of Amadeo Rea*, edited by Marsha Quinlan and Dana Lepofsky (Tacoma, WA: Society of Ethnobiology, 2012). Pay for and download this particular essay as a PDF at http://ethnobiology.org/13-feed-all-people-lucille-clifton-s-fall-feasts-gitga-community-hartley-bay-british-columbia.

588 International Boreal Conservation Science Panel and Associates, "Conserving the World's Last Great Forest Is Possible: Here's How," Boreal Science, July 2013, http://borealscience.org/wp-content/uploads/2013/07/conserving-last-great-forests1.pdf.

589 Christopher M. Raymond, Gerald Singh, Karina Benessaiah, Joanna R. Bernhardt, Jordan Levine, Harry Nelson, Nancy J. Turner, Bryan Norton, Jordan Tam and Kai M.A. Chan, "Ecosystem Services and Beyond: Using Multiple Metaphors to Understand Human-Environment Relationships," *BioScience* 63, no. 7 (2013): 536–46, http://bioscience.oxfordjournals.org/content/63/7/536.

590 Nancy Turner, *Ancient Pathways, Ancestral Knowledge: Ethnobotany and Ecological Wisdom of Indigenous Peoples of Northwestern North America*, 2 vols. (Montreal/Kingston: McGill-Queen's University Press, 2014). For a review of the first volume, see Amy Reiswig, "*Ancient Pathways, Ancestral Knowledge*: Nancy Turner's Monumental New Work Explores Humanity's Multi-Faceted Relationship to Plants," *Focus Online*, June 2014, http://www.focusonline.ca/?q=node/729.

591 "Two More Faculty Join Distinguished Professorship Ranks," *The Ring* 30, no. 2 (February 2004), http://ring.uvic.ca/04feb04/features/dist-prof.html.

592 Some of her awards include the University of Victoria Craigdarroch Gold Medal for Career Achievement in Research, Academic of the Year from the Confederation of University Faculty Associations of B.C., the Order of British Columbia and the R.E. Schultes Award from the U.S.-based Healing Forest Conservancy, and honorary doctorates from Vancouver Island University and University of British Columbia. See "UVic

Ethnobotanist Named to Order of Canada," University of Victoria, July 17, 2009, http://communications.uvic.ca/uvicinfo/announcement.php?id=342. See also "Environmental Studies, Nancy Turner," University of Victoria, accessed March 16, 2015, http://www.uvic.ca/socialsciences/environmental/people/faculty/turnernancy.php.

593 "The Land of the Secwepemc," Land of the Shuswap, accessed March 16, 2015, http://www.landoftheshuswap.com/land.html.

594 Jim Cooperman, "The Rich Legacy of Dr. Mary Thomas," *Shuswap Passion* (blog) May 2, 2014, http://shuswappassion.ca/history/the-rich-legacy-of-dr-mary-thomas/.

595 Ibid. See also the Mary Thomas Cultural Centre and Heritage Sanctuary, http://www.shuswapcentre.org/#.

596 Her awards include honorary degrees from the University of Victoria and the University of North Carolina. She became the first Native Canadian recipient of the the Seacology Foundation's Indigenous Conservationist Award. Dedicated also to improving the lives of young people, Mary earned the Medal for Exceptional Contributions to Early Childhood Development from the Centre of Excellence for Early Childhood Development. Two Mary Thomas Scholarships are awarded to the two top Aboriginal early childhood education students each year. See Cooperman, "The Rich Legacy of Dr. Mary Thomas."

597 Hannah Askew, "Water Fight in the Thompson Okanagan, First Nations Struggle to Take Back Their Water Resources," *Briarpatch Magazine*, January 5, 2010, http://briarpatchmagazine.com/articles/view/water-fight-in-the-thompson-okanagan. See also Lyse C. Cantin, "Wetlands Enhancement Project: A Model for Environmental Cooperation," Wetland Alliance (WA:TER), accessed March 15, 2015, http://www.water.ca/MaryThomas.pdf.

598 See T. Abe Lloyd, *Wild Harvests: Wild Food Experiments and Personal Foraging Accounts from the Pacific Northwest Centering on Northwest Washington and Southern Vancouver Island* (blog), accessed March 16, 2015, http://arcadianabe.blogspot.ca/p/foraging-courses-instructors-and-blogs.html. See also T. Abe Lloyd, "Wapato: Cultivating Native Tubers," *Wild Harvests* (blog), October 8, 2012, http://arcadianabe.blogspot.ca/2012/10/wapato-cultivating-native-tubers.html.

599 Ibid.

600 "Meet Elizabeth May," Elizabeth May, MP, accessed July 12, 2014, Elizabethmaymp.ca/home/meet-elizabeth-widget/meet-elizabeth-may.

601 Elizabeth May, *Who We*

Are: Reflections on My Life and Canada (Vancouver: Greystone Books, 2012).

602 See http://www. elizabethmaymp.ca/

603 See www. chefsurvivalchallenge.com.

604 "Biography," David Suzuki Foundation, accessed July 12, 2014, http://www.davidsuzuki. org/david/.

605 *The Nature of Things*, CBC, accessed July 12, 2014, http://www.cbc.ca/ natureofthings/.

606 "Critical Issues Deserve a Higher Standard," David Suzuki Foundation, October 31, 2013, http://davidsuzuki. org/blogs/science-matters/2013/10/critical-issues-deserve-a-higher-standard.

607 "We Can't Ignore the Little Things That Keep Us Alive," David Suzuki Foundation, August 15, 2013, http://www. davidsuzuki.org/blogs/ science-matters/2013/08/ we-cant-ignore-the-little-things-that-keep-us-alive/. "The small things that keep the biosphere going for creatures like us are probably more threatened because we ignore them. If we spend time studying them, they have much to teach us."

608 David Suzuki, "Small Farms May Be Better for Food Security and Biodiversity," *Straight*, June 14, 2011, http:// www.straight.com/news/ david-suzuki-small-farms-may-be-better-food-security-and-biodiversity. See also

Michael Jahi Chappell and Liliana A. LaValle, "Food Security and Biodiversity: Can We Have Both? An Agroecological Analysis," *Agriculture and Human Values* 28, no. 1 (February 2011): 3–26.

609 On May 4, 2011, HRH Charles, The Prince of Wales, gave the keynote speech at the Future of Food conference at Georgetown University in Washington, DC. To hear his 50-minute speech, go to "The Speech," On the Future of Food, accessed July 12, 2014, http://www. onthefutureoffood.org/ the-speech.

610 "Eco-Farming Can Double Food Production in 10 Years, Says New UN Report," United Nations Human Rights, News Release, March 8, 2011, http://www.srfood. org/images/stories/pdf/ press_releases/20110308_ agroecology-report-pr_en.pdf.

611 "International Sustainability Unit," The Prince's Charities, accessed July 12, 2014, http:// www.pcfisu.org/.

612 The Prince of Wales, "Reward Good Food: Prince Charles on Healthy, Sustainable Farming," *The Atlantic*, February 22, 2012, http://www.theatlantic. com/health/archive/2012/02/ reward-good-food-prince-charles-on-healthy-sustainable-farming/253350/. To hear his 50-minute speech, go to "The Speech," On the Future of Food. See also HRH, The Prince of Wales, *The Prince's Speech: On the Future of Food* (Emmaus: Rodale

Books, 2012).

613 "Prince Charles – The European Nature Trust (TENT): Living in Harmony with Nature in Romania," in *Wild Carpathia, Part One,* YouTube video, 5:55, posted by "UbuntuLiving," May 23, 2012, http://www.youtube.com/watch?v=w1tVEpQmWW4.

614 HRH, The Prince of Wales, *The Prince's Speech.*

615 Vandana Shiva, *Hidden Variables and Locality in Quantum Theory* (London, ON: Faculty of Graduate Studies, University of Western Ontario, 1978), http://amicus.collectionscanada.ca/aaweb-bin/aamain/itemdisp?sessionKey=999999999_142&l=0&d=2&v=0&lvl=1&itm=3397435.

616 "Archive for Vandana Shiva," Global Alliance for the Rights of Nature, accessed July 12, 2014, https://therightsofnature.org/tag/vandana-shiva/.

617 "Rights of Mother Earth," World People's Conference on Climate Change and the Rights of Mother Earth, accessed July 12 2014, http://pwccc.wordpress.com/programa/.

618 "Universal Declaration of Rights of Mother Earth," Global Alliance for the Rights of Nature, accessed July 12, 2014, http://therightsofnature.org/universal-declaration/.

619 "Dr Vandana Shiva Rights of Nature Tribunal President Statement," YouTube video, 36:30, posted by "Rights4Nature," February 5, 2014, https://www.youtube.com/watch?v=k3Tlo2_bcbw.

620 Ibid.

621 Ibid.

622 Vandana Shiva, Debbie Barker and Caroline Lockhart, coordinators, *The GMO Emperor Has No Clothes: A Global Citizens Report on the State of GMOs – False Promises, Failed Technologies,* accessed June 18, 2014, http://www.navdanya.org/attachments/Latest_Publications9.pdf.

623 See "Nutritional Info of: Lambsquarters, raw, vs. Spinach, raw," Skip the Pie, accessed July 12, 2014, http://skipthepie.org/vegetables-and-vegetable-products/lambsquarters-raw/compared-to/spinach-raw/#nut.

624 "Acceptance Speech by Vandana Shiva," The Right Livelihood Award, December 9, 1993, http://www.rightlivelihood.org/shiva_speech.html.

625 "The Green Revolution," Ecotextiles, September 6, 2001, http://oecotextiles.wordpress.com/2011/06/09/the-green-revolution/.

626 Shiva, Barkert and Lockhart, *The GMO Emperor,* 21. "Patents provide royalties for the patent holder and corporate monopolies. This translates into super profits for Monsanto. For the farmers this means debt. For example, more than 250,000 Indian farmers have been pushed to suicide in the last decade and

a half. Most of the suicides are in the cotton belt where Monsanto has established a seed monopoly through Bt cotton."

627 Vandana Shiva, *Earth Democracy: Justice, Sustainability, and Peace* (Cambridge, Mass.: South End Press, 2005), 67–68.

628 "Bija Vidyapeeth – Earth University," Navdanya, accessed July 12, 2014, http://www.navdanya.org/earth-university.

NATHALIE CHAMBERS has a diploma in Restoration Ecology from the University of Victoria and has studied Conservation Finance at Yale University. She is the founder of both the Chef Survival Challenge Inc., a fundraising event that channels proceeds to farmland conservation, and the Big Dream Farm Fund, which directs funds towards farmland acquisition and sustainable farming education initiatives. Nathalie and her husband, David Chambers, live and work on Madrona Farm, where they grow more than a hundred varieties of produce year-round for over 4,000 regular customers, including numerous wholesalers and local restaurants. Nathalie is mother to two kids, Sage and Lola, and lives in Victoria, BC.

ROBIN ALYS ROBERTS, B.Ed. with Applied Linguistics, taught at the University of Victoria. She has written a tutorial manual, websites, newsletters, conference presentations and magazine articles; co-authored yachting books; and edited theses, résumés, websites and a peace education book. A happy grandmother, she loves gardening, good food and democracy. She lives in Victoria, BC.

SOPHIE WOODING has a degree in creative writing and English literature from the University of Victoria. She apprenticed at Madrona Farm surrounding a stint at GoodRoots, a Community Sustained Agriculture Farm near her hometown of Langley, BC. Sophie lives in Victoria, BC.